Essential Fatty Acids

Essential Fatty Acids

Sources, Processing Effects, and Health Benefits

Edited by
Sanju Bala Dhull, Sneh Punia,
Kawaljit Singh Sandhu

CRC Press
Taylor & Francis Group
Boca Raton London New York

CRC Press is an imprint of the
Taylor & Francis Group, an **informa** business

First edition published 2021
by CRC Press
6000 Broken Sound Parkway NW, Suite 300, Boca Raton, FL 33487-2742

and by CRC Press
2 Park Square, Milton Park, Abingdon, Oxon, OX14 4RN

Library of Congress Cataloging-in-Publication Data

Names: Dhull, Sanju Bala, editor. | Punia, Sneh, editor. | Sandhu, Kawaljit Singh, editor.
Title: Essential fatty acids : sources, processing effects, and health benefits / edited by Sanju Bala Dhull, Sneh Punia, Kawaljit Singh Sandhu.
Description: First edition. | Boca Raton : CRC, 2021. | Includes bibliographical references and index.
Identifiers: LCCN 2020028959 (print) | LCCN 2020028960 (ebook) | ISBN 9780367335403 (hardback) | ISBN 9780429321115 (ebook)
Subjects: LCSH: Fatty acids in human nutrition. | Fatty acids--Health aspects. | Fatty acids--Physiological effect. | Essential fatty acids.
Classification: LCC QP752.F35 E87 2021 (print) | LCC QP752.F35 (ebook) | DDC 612.3/97--dc23
LC record available at https://lccn.loc.gov/2020028959
LC ebook record available at https://lccn.loc.gov/2020028960

ISBN: 978-0-367-33540-3 (hbk)
ISBN: 978-0-429-32111-5 (ebk)

Typeset in Palatino
by Deanta Global Publishing Services, Chennai, India

CONTENTS

PREFACE

Essential fatty acids (α-linolenic acid and linoleic acid) are required for good health but cannot be synthesized by the body. Therefore, these nutrients must be supplied from either diet or dietary supplements. A balanced ω-6:ω-3 ratio is desirable; therefore, a diet rich in ω-6 or poor in ω-3 might result in various diseases. Recent studies have increased the scientific and medical interest because of the beneficial effects of these fatty acids on brain and retina function, as well as reducing ill health effects such as cardio-metabolic diseases. Therefore, interest in their requirements and dietary recommendations has also recently increased. This book will explore the knowledge about the benefits, structure, chemical composition and intake recommendations for essential fatty acids.

Chapter 1 deals with the basic composition, structure and types of fatty acids. In this chapter, saturated and unsaturated fatty acids and their different characteristics are discussed. Essential fatty acids are also elaborated in this chapter. Essential fatty acids are lipids that are not metabolized by the body but are involved in various biological processes, producing many compounds when they are metabolized. These fatty acids play an important role in the prevention of various diseases.

Chapter 2 explores the different sources of essential fatty acids with their respective importance. Essential fatty acids can be obtained from plant as well as animal sources. Among plant sources, flaxseed oil is the richest source of essential fatty acids. The main sources of essential fatty acids are animal and marine sources. Due to many issues regarding the production of essential fatty acids, the main concern of research is genetically engineered and microbial sources.

Chapter 3 discusses the chemistry of the essential fatty acids, including their chemical structure; their chemistry is vital in terms of their applications. In this chapter, the nomenclature of essential amino acids is also discussed. Essential fatty acids consist of double bonds in their aliphatic chain system; according to the IUPAC system, the *anoic* is transformed into the *enoic* to enable the occurrence of a double bond. As well, with the number of double bonds in essential fatty acids, the terms *di-*, *tri-*, *tetra-* and so on are introduced. Structures of omega-3, omega-6 and omega-9 are elaborately discussed in this chapter.

Chapter 4 explores the physical, chemical and functional properties of fatty acids. Fatty acids are classified in terms of the presence and number or quantity of double bonds as saturated and unsaturated fatty acids. Saturated fatty acids are straight hydrocarbon chains with even numbers of carbon atoms, while unsaturated fatty acids have at the minimum one double bond incorporated in their hydrocarbon chains. Physical properties such as melting point, oxidation stability, smoke point, solubility, specific heat and specific gravity are discussed in this chapter. Chemical and functional properties are also elaborated.

Chapter 5 deals with metabolism of essential fatty acids. In this chapter, the metabolism of fatty acids, especially α-linolenic acid, linoleic acid and arachidonic acid, is discussed. Dietary patterns, including consumption of essential fatty acids, are positively correlated with reducing morbidity and mortality from cancer, hypertension, and cardiovascular diseases; anti-inflammatory and anti-carcinogenic potentials are also discussed.

Chapter 6 focuses on the effect of thermal and non-thermal processing on essential fatty acids. Essential polyunsaturated fatty acids omega-3 and -6 are highly susceptible substrates during thermal processing, which might result in their potential loss quantitatively as well as qualitatively. Non-thermal processes—i.e. micro- and nano-encapsulation; use of antioxidants, surfactants and emulsifiers; and cold plasma and pulsed electric field—are elaborated in this chapter. Non-thermal technologies have recently emerged as alternatives over thermal treatment for processing polyunsaturated fatty acid–enriched foods.

Chapter 7 deals with the effect of storage on essential fatty acids from marine, egg, vegetable oil and milk sources. The storage stability of fatty acids under variable storage conditions is discussed in this chapter. Storage stability is influenced by various factors such as the composition of fatty acids in products, level of pro-oxidants, enzymes, iron (Fe^{+2}) concentration, storage temperature, pH, oxygen level and ionic strength. Different storage conditions are used to improve the shelf life of oil-based food products.

Chapter 8 explores enhancements in the bio-availability of essential fatty acids. Many techniques such as self-micro-emulsifying delivery system, encapsulation, emulsification, and alpha-tocopheryl phosphate mixture are used to enhance the bio-availability of essential fatty acids. Research is conducted to improve the bio-availability of milk, meat, algae, poultry and many more food products.

Enrichment of essential fatty acids is discussed in Chapter 9. As humans cannot synthesize essential fatty acids, it is necessary to enrich our diet with essential fatty acids via external sources. The major sources of all the essential fatty acids are mentioned in the chapter. Walnuts, flaxseeds, canola, fish oil, seafood and flaxseed oil are normally used for enrichment of food with fatty acids. Many processing techniques are discussed to improve the essential fatty acid profile of foods.

Dietary intake and recommendations for essential amino acids are elaborated in Chapter 10. This chapter broadly describes the fat and essential fatty acid intake recommendations for different age groups. The primary focus of this chapter is on the total intake and composition of essential fatty acids and the dietary lipid supply as a major determinant of growth and long-term health. Historical changes in dietary patterns and the ω-6/ω-3 ratio for optimal human health are also discussed in this chapter.

Chapter 11 discusses the health benefits of essential fatty acids. Linoleic and α-linolenic acid are essential as they must be supplied in the diet because of their major roles in the human body. Essential fatty acids are vital in prevention of inflammation, arthritis, cardiovascular diseases, diabetes, mental health problems, Alzheimer's disease and cancer. Microencapsulated foods are the future foods for the supplementation of essential fatty acids.

Chapter 12 deals with the health benefits and future prospects of dietary fatty acids. In this chapter, different techniques to preserve essential fatty acids are discussed. Tailor-made fats (also known as structured lipids) are triacylglycerols that have been made by the incorporation of new fatty acids and restructured to change the positions of fatty acids. These fatty acids are made to achieve various health benefits in the human body. Toxicity and safety issues regarding fatty acids are also elaborated in this chapter.

This book is useful for students, academicians, researchers and other interested professionals working on the metabolism, health benefits and other properties of essential fatty acids. There are many books available on fatty acids, but this book is designed in such a way that it deals with important aspects related to essential fatty acids. The authors appreciate further comments/information for future reference.

EDITORS

Sanju Bala Dhull, PhD is a Senior Assistant Professor at the Department of Food Science and Technology, Chaudhary Devi Lal University, Sirsa, and has more than 12 years of teaching and research experience. Her areas of interest include synthesis and characterization of nanoparticles, nanoemulsions and antimicrobial activity of nanoformulations in food, characterization of biomolecules, starch and antioxidants. She has published more than 25 research papers in journals of national and international repute. She has presented more than 20 research papers at various national and international conferences. She is an active member of The Association of Food Scientists and Technologists (India) and The Association of Microbiologists of India. She also serves as a reviewer for several national and international journals.

Sneh Punia, PhD is an Assistant Professor at the Department of Food Science and Technology, CDLU, Sirsa. Her areas of interest include cereal antioxidants, starch, and the development of new products. She has published more than 35 works, including refereed publications, an authored book, an edited book, as well as book chapters and communications at national and international conferences. She also serves as the reviewer for various international journals. She is an active member of the Association of Food Scientists and Technologists (India).

Kawaljit Singh Sandhu, PhD is an Associate Professor at the Department of Food Science and Technology, Maharaja Ranjeet Singh Punjab Technical University, Bathinda. He was awarded a postdoctoral fellowship from Korea University, Seoul, South Korea. With more than 15 years of research and teaching experience, his research interest is focused on starch, starch modification, antioxidants, nano-technology and drug delivery. Dr. Sandhu received several research projects from UGC, New Delhi and DST, New Delhi. He was awarded the Young Scientist Award from the Association of Food Scientists and Technologists (AFSTI), Mysore, India. Dr. Sandhu is an active member of AFSTI, India, the Association of Microbiologists of India and the Korean Society of Food Science and Technology, South Korea. He has published more than 50 research papers in various national and international journals.

CONTRIBUTORS

Aarti Bains
Chandigarh Group of Colleges
Mohali, India

Aradhita Barmanray
Department of Food Technology
Guru Jambheshwar University of Science and Technology
Hisar, India

Vandana Chaudhary
College of Dairy Science and Technology
Lala Lajpat Rai University of Veterinary and Animal Sciences
Hisar, India

Prince Chawla
Department of Food Technology and Nutrition
Lovely Professional University
Phagwara, India

Rohit Chhetri
School of Bioengineering and Food Technology
Shoolini University
Solan, India

Amanjyoti Gill
Department of Food Science and Technology
Chaudhary Devi Lal University
Sirsa, India

Priyanka Kajla
Department of Food Technology
Guru Jambheshwar University of Science and Technology
Hisar, India

Ramandeep Kaur
Department of Food Science & Technology
Punjab Agricultural University
Ludhiana, India

Ravinder Kaushik
Amity Institute of Food Technology
Amity University
Noida, India

Mohammad Kashif Kidwai
Department of Energy and Environmental Science
Chaudhary Devi Lal University
Sirsa, India

Pradyuman Kumar
Department of Food Engineering & Technology, SLIET
Longowal, India

Parveen Kumari
Department of Food Technology
Guru Jambheshwar University of Science and Technology
Hisar, India

Anju Malik
Department of Energy and Environmental Science
Chaudhary Devi Lal University
Sirsa, India

Tanu Malik
Centre of Food Science and Technology
CCS Haryana Agriculture University
Hisar, India

Mohd. Mohsin
Department of Biosciences
Jamia Millia Islamia
New Delhi, India

Kiran Bala Nain
Department of Food Science and Technology
University College, Chaudhary Devi Lal University
Sirsa, India

Dipsha Narang
Maharaja Lakshman Sen Memorial College
Sundernagar, India

Manju Nehra
Department of Food Science and Technology
Chaudhary Devi Lal University
Sirsa, India

Charanjiv Singh Sain
Department of Food Engineering and Technology
Sant Longowal Institute of Engineering and Technology
Longowal, India

Loveleen Sharma
Amity Institute of Food Technology
Amity University
Noida, India

Ajay Singh
Department of Food Technology
Mata Gujri College
Fatehgarh Sahib, India

Ashish Kumar Singh
Dairy Technology Division
National Dairy Research Institute
Karnal, India

Anil Kumar Siroha
Department of Food Science and Technology
Chaudhary Devi Lal University
Sirsa, India

Mudasir Yaqoob
Department of Food Technology and Nutrition
Lovely Professional University
Phagwara, India

LIST OF ABBREVIATIONS

4-HHE	4-Hydroxy-2-hexenal
AA	Arachidonic acid
ACBP	Acyl-CoA-binding protein
ADHD	Attention deficit hyperactivity disorder
ALA	α-Linolenic acid
ALT	advanced lipid technologies
AMDR	Acceptable macronutrient distribution range
ARA	Arachidonic acid
AV	Anisidine value
CHD	Coronary heart disease
CLA	Conjugated linoleic acid
CNS	Central nervous system
CO_2	Carbon dioxide
COPs	Cholesterol oxidation products
COX	Cyclooxygenase
CPS	Canadian Pediatric Society
CS	Chitosan
CSM	Cress seed mucilage
Cu	Copper
DAG	Di-acyl glycerol
DDM	Dialkyl dihexadecylmalonate
DGLA	Dihomo-gamma-linolenic acid
DHA	Docosahexaenoic acid
DPA	Docosapentaenoic acid
EE	Ethyl ester
EFA	Essential fatty acids
EPA	Eicosapentaenoic acid
EPA	Environmental Protection Agency
EPGs	Esterified propoxylated glycerols
ESPGHAN	European Society for Paediatric Gastrology, Hepatology, and Nutrition
ETA	Eicosatetraenoic acid
EU	European Union
F3-IsoPs	F3-isoprostans
FA(s)	fatty acid(s)

FAO	Food and Agriculture Organization
FATF	Fatty acid transport protein
FDA	Food and Drug Administration
Fe^{+2}	Ferrous ions
Fe	Iron
FFA	Free fatty acid
GLA	Gamma-linolenic acid
GM	Genetically modified
GOED	Global Organization for EPA and DHA Omega-3s
GRAS	Generally Recognized as Safe
GSH	Glutathione
HDL	High-density lipoproteins
HGL	Human gastric lipase
HHE	4-Hydroxy-2-hexenal
HPL	Human pancreatic lipase
HUFA	Highly unsaturated
IUPAC	International Union of Pure and Applied Chemistry
IV	Iodine value
LA	Linoleic acid
LC ω3 FAs	Long-chain omega-3 fatty acids
LC-HUFA	Long-chain highly unsaturated fatty acids
LC	Long chain
LC-PUFAs	Long-chain polyunsaturated fatty acids
LDL	Low-density lipoprotein
MAG	2-Monoacylglycerol
MDA	Malondialdehyde
meq	Milliequivalents
MUFA(s)	Monounsaturated fatty acid(s)
n-3	Omega-3
n-6	Omega-6
NEFAs	Non-essential fatty acids
Nrf2	Nuclear factor erythroid 2-related factor2
PC	Phosphatidylcholine
PCOD	Polycystic ovarian disorder
PEF	Pulsed electric field
PET	Polyethylene terephthalate
PL	Plasma lipid
PUFA(s)	Polyunsaturated fatty acid(s)
PV	Peroxide value
RI	Refractive index

SCF	The European Scientific Committee on Food
SDA	Stearidonic acid
SEDDS	Self-emulsifying drug delivery system
SFA(s)	Saturated fatty acid(s)
SMEDS	Self-microemulsifying delivery system
TAC	Trialkoxy citrate
TAGs	Triacylglycerols
TATCA	Trialkoxy tricarballylate
TBA	Thiobarbituric acid
TBARS	Thiobarbituric acid reactive substances
TFAs	Trans fatty acids
TG	Triglyceride
TGE	Trialkoxy glyceryl ether
TNF-α	Tumour necrosis factor-alpha
TPA	Tetracosapentaenoic acid
UFAs	Unsaturated fatty acids
VLDL	Very-low-density lipoprotein
WHO	World Health Organization
WPI-GA	Whey protein isolate-gum arabic
w/w	weight/weight
YTA	Yetracosahexaenoic acid
Ω	Omega

1

Essential Fatty Acids
Introduction

Sanju Bala Dhull and Sneh Punia

Contents

1.1 INTRODUCTION

Fats are an essential part of our life, provide a major source of fuel for the body and help in the absorption of fat-soluble vitamins. A fat molecule has a glycerol backbone and three fatty acid tails. Glycerol is a small organic molecule with three hydroxyl (OH) groups, while a fatty acid (FA) consists of a long hydrocarbon chain attached to a carboxyl group. Fatty acids, both free and as a part of complex lipids, play an important role in normal operation of body organ systems. They serve as a metabolic fuel system and as components of cell membranes and act as gene regulators (Rustan and Drevon, 2005). For proper functioning of the respiratory system, circulatory system, nervous system, immune system, etc., the body requires fatty acids (Kapalka, 2009). A fatty acid is an organic acid that has an acid group at one end of its molecule and a methyl group at the other end (Whitney et al., 1998). There are many different fatty acid species on the presence of double bond, i.e., 0 (saturated), 1 (monounsaturated), >1 (PUFA) and >2 (highly unsaturated, HUFA).

1.2 HISTORY

Approximately 100 years ago, a major question in biochemistry was whether a fat was to be considered essential or not, which seems quite simple in this era of genomics, proteomics and lipidomics. Earlier, in the first part of the 20th century, dietary fat was recognized as a good source of energy and fat-soluble vitamins. As the fatty acid could be synthesized from carbohydrate supplied by the diet, the essentiality of fatty acids was not considered. In 1918, the first indication about the essentiality of fat for health was presented by Aron by suggesting that butter has a specific nutrient value that other dietary components cannot provide (Aron, 1919). His study reported that apart from calorific value, butter fat had some inherent special nutritive value which might be due to a certain lipid in it. In 1929, a young plant physiology assistant professor at the University of Minnesota, George Oswald Burr, initially reported deficiency disease in rats fed on a fat-free diet and concluded that fat was an essential dietary component (Burr and Burr, 1929). The study concluded that warm-blooded animals cannot synthesize sufficient amounts of some fatty acids. In 1930, Burr and Burr further added to their work by concluding that linoleic acid (LA or 18:2n-6) inclusion in the diet can reverse all the deficiency symptoms seen in their earlier study (Burr and Burr, 1929). Thus, LA was heralded as an essential fatty acid (EFA) (Burr and Burr, 1930). The recognition that the human body cannot synthesize some unsaturated FAs (UFAs) and needs dietary supplementation led to the categorization of essential and non-essential FAs. Originally, only two FAs were considered as essential: linoleic acid (9,12-octadecadienoic acid, LA, 18:2n-6) and α-linolenic acid (9,12,15-octadecatrienoic acid, ALA, 18:3n-3). However, continuous further nutritional and clinical studies produced strong evidence not only in support of LA and ALA, but also for arachidonic acid and other long-chain highly unsaturated fatty acids (LC-HUFA) such as eicosapentaenoic acid (C20:5n-3, EPA) and docosahexaenoic acid (C22:6n-3, DHA) (Turpeinen, 1937; Burr et al., 1940; Quackenbush et al., 1942). More recent studies on the health benefits and biological significance of these LC n-3 and n-6 polyunsaturated FAs (PUFAs) do qualify them as EFAs.

1.3 OVERVIEW OF FATTY ACID STRUCTURE

A fatty acid is a carboxylic acid with a long side chain of hydrocarbons. Most fatty acids contain an even number of carbon atoms in the hydrocarbon chain and follow the general molecular formula of $CH_3(CH_2)_n$ COOH, with the methyl/omega (ω) at one end and carboxyl at the other end. The carbon atom next to the carboxyl group is called the α carbon, and the subsequent one the β carbon (Figure 1.1). The letter n is also often used to describe the position of double bonds closest to the methyl end.

1.4 TYPES OF FATTY ACIDS

Fatty acids are further categorized into saturated and unsaturated fatty acids (Figure 1.2). Saturated fatty acids are called so as they are saturated or filled with hydrogen atoms (Figure 1.3a). Generally they are straight hydrocarbon chains with even number of carbon atoms (12–22 carbon atoms). UFAs are typically biosynthesized by a unique dehydrogenation reaction called desaturation (Buist, 2010) and contain one or more double/triple carbon–carbon bonds in the carbon chain. On the basis of bonds, unsaturated fatty acids are of two types. Monounsaturated fatty acids have only one (C=C) double bond with *cis*-configuration in a chain of 16–22 carbon atoms, which can occur in different positions (Figure 1.3b). In polyunsaturated fatty acids, the first double bond may be found between the third and the fourth carbon atom from the carboxyl group; these are called ω-3

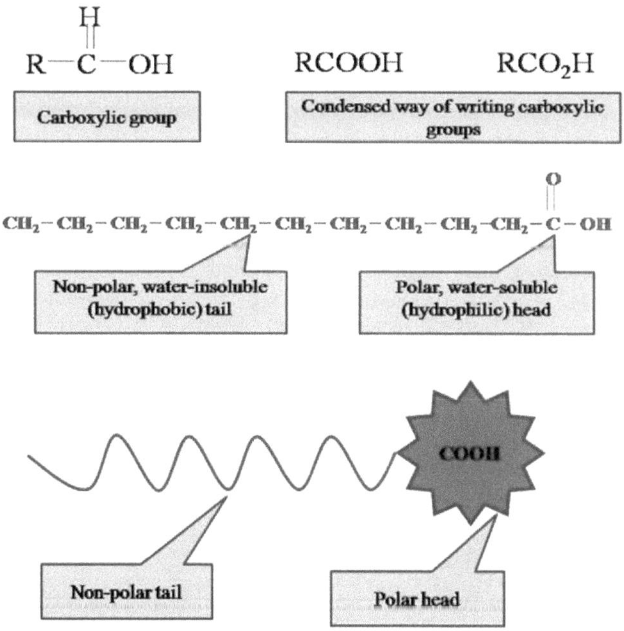

Figure 1.1 General molecular formula of fatty acids

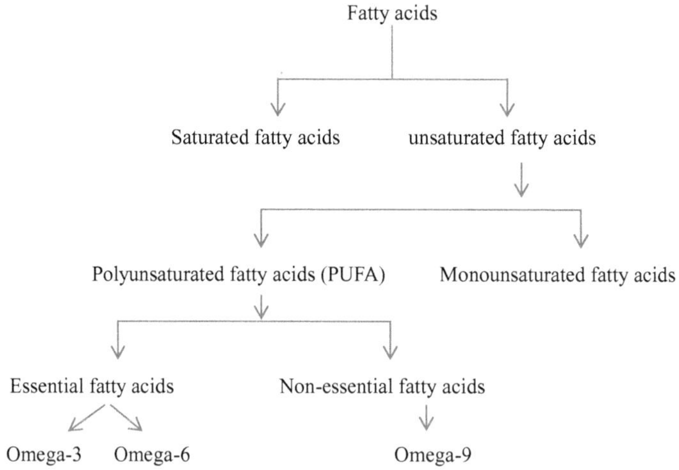

Figure 1.2 General fatty acid classification

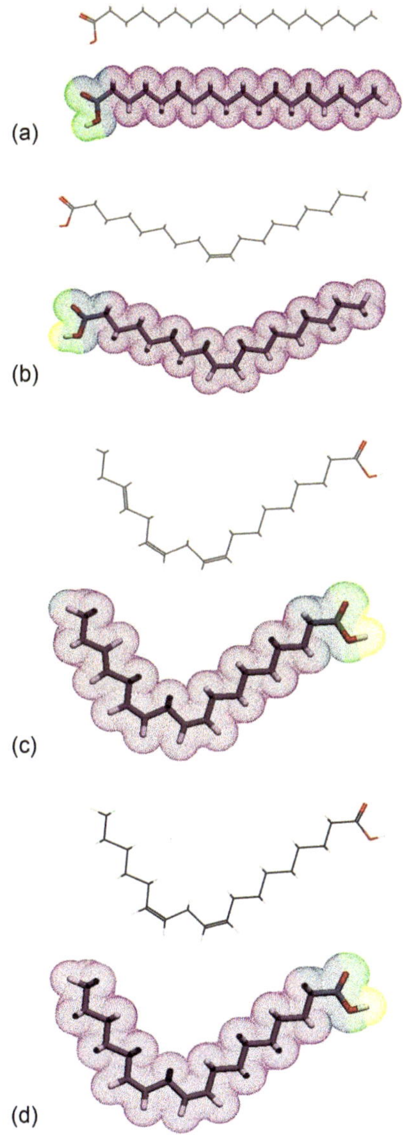

Figure 1.3 (a) A saturated fatty acid (stearic acid); (b) a monounsaturated fatty acid (oleic acid); (c) an omega-3 polyunsaturated fatty acid (linolenic acid); (d) an omega-6 polyunsaturated fatty acid (linoleic acid)

fatty acids (Figure 1.3c). If the first double bond is between the sixth and seventh carbon atoms, then they are called ω-6 fatty acids (Figure 1.3d). The double bonds in PUFAs are separated from each other by methylene grouping. An inflexible "kink" inserted by the *cis*-double bonds in the carbon chain of UFAs prevents their tight packing as that of saturated fatty acids, resulting in a lower melting point in comparison with saturated fats. The general structural formula of fatty acids for saturated fatty acids is $CnH_{2n}O_2$, for monounsaturated fatty acids it is $CnH_{2n-2}O_2$, and for polyunsaturated fatty acids it is $CnH_{2n2x}O_2$ where x = number of double bonds (Table 1.1). The melting points of fatty acids also greatly vary with number of carbon atoms; shorter fatty acids (e.g., palmitic acid, 16C, melting point = 63°C) generally have lower melting points compared with longer ones (e.g., stearic acid, 18C, melting point = 70°C). Of total energy intake, fatty acids represent 30–35% and vegetable oils, dairy products, meat products, fish oils, etc., are the dietary sources of fatty acids.

1.4.1 Saturated Fatty Acids

The most common saturated fatty acid in animals, plants and microorganisms is palmitic acid (16:0). Stearic acid (18:0, Figure 1.3a) is a major saturated fatty acid in animals and some fungi, and a minor component in most plants. Myristic acid (14:0) has a widespread occurrence, occasionally as a major component. Shorter-chain saturated acids with 8–10 carbon atoms are found in milk and coconut triacylglycerols. A list of saturated fatty acids is shown in Table 1.1. Many epidemiological studies have reported that consumption of long-chain saturated fatty acids (12–18 carbon atoms chain) is associated with cardiovascular diseases. Eating foods that contain saturated fats raises the level of cholesterol in blood. High levels of LDL cholesterol in blood increase risk of heart disease and stroke.

1.4.2 Unsaturated Fatty Acids

There has been significant consumer interest in the health-enhancing role of specific foods or physiologically active food components. Therefore, nowadays people have focused their interest in consuming unsaturated fats rather than saturated fats. Considering health as a priority, people have become very aware of the health benefits and disease prevention by increasing the consumption of polyunsaturated/essential fatty acids (Punia et al., 2019). It has been reported that replacement of saturated fatty

Table 1.1 A List of Saturated and Unsaturated Fatty Acids

Common Name	IUPAC Nomenclature	Condensed Structure Formula	Abbreviated Structural Formula
Saturated fatty acids			
Butyric acid	Butanoic acid	$CH_3(CH_2)_2COOH$	C_3H_7COOH
Caproic	Hexanoic acid	$CH_3(CH_2)_4COOH$	$C_5H_{11}COOH$
Caprylic	Octanoic acid	$CH_3(CH_2)_6COOH$	$C_7H_{15}COOH$
Capric	Decanoic acid	$CH_3(CH_2)_8COOH$	$C_9H_{19}COOH$
Lauric	Dodecanoic acid	$CH_3(CH_2)_{10}COOH$	$C_{11}H_{23}COOH$
Myristic	Tetradecanoic acid	$CH_3(CH_2)_{12}COOH$	$C_{13}H_{27}COOH$
Palmitic	Hexadecanoic acid	$CH_3(CH_2)_{14}COOH$	$C_{15}H_{31}COOH$
Stearic	Octadecanoic acid	$CH_3(CH_2)_{16}COOH$	$C_{17}H_{35}COOH$
Arachidic	Eicosanoic acid	$CH_3(CH_2)_{18}COOH$	$C_{19}H_{39}COOH$
Behenic	Docosanoic acid	$CH_3(CH_2)_{20}COOH$	$C_{21}H_{44}COOH$
Unsaturated fatty acids			
Oleic acid	Octadec-2-enoic acid	$CH_3(CH_2)_7CH{=}CH(CH_2)_7COOH$	$C_{17}H_{33}COOH$
Linoleic acid	(9Z,12Z)-octadeca-9,12-dienoic acid	$CH_3(CH_2)_4CH{=}CHCH_2CH{=}CH(CH_2)_7COOH$	$C_{17}H_{31}COOH$
Alpha-linolenic acid	(9Z,12Z,15Z)-octadeca-9,12,15-trienoic acid	$CH_3CH_2CH{=}CHCH_2CH{=}CHCH_2CH{=}CH(CH_2)_7COOH$	$C_{17}H_{29}COOH$

(*Continued*)

Table 1.1 (Continued) A List of Saturated and Unsaturated Fatty Acids

Common Name	IUPAC Nomenclature	Condensed Structure Formula	Abbreviated Structural Formula
Arachidonic acid	(5Z,8Z,11Z,14Z)-icosa-5,8,11,14-tetraenoic acid	$CH_3(CH_2)_4CH{=}CHCH_2CH{=}CHCH_2CH{=}CHCH_2CH{=}CH(CH_2)_3COOH$	$C_{19}H_{31}COOH$
Eicosapentaenoic acid	(5Z,8Z,11Z,14Z,17Z)-icosa-5,8,11,14,17-pentaenoic acid	C20:5	$C_{19}H_{29}COOH$
Docosahexaenoic acid	(7Z,10Z,13Z,16Z,19Z)-docosa-7,10,13,16,19-pentaenoic acid	C22:6	$C_{21}H_{31}COOH$
Erucic acid	(Z)-Docos-13-enoic acid	$CH_3(CH_2)_7CH{=}CH(CH_2)_{11}COOH$	$C_{21}H_{41}COOH$

acids by polyunsaturated fatty acids resulted in a 17% lower risk of cardiovascular events (Jakobsen et al., 2009).

1.4.2.1 Monounsaturated Fatty Acids

Humans have the potential to synthesize saturated fatty acids and some monounsaturated fatty acids (oleic acid (18:1 ω-9), most common monoenoic fatty acid in plants and animals) from the carbon group present in proteins and carbohydrates. But omega-3 and omega-6 fatty acids cannot be synthesized in the human body because of the lack of enzymes necessary to insert a *cis*-double bond at the n-3 and n-6 positions of the carbon chain, making them essential fatty acids (Food and Nutrition Board, 2002).

1.4.2.2 Polyunsaturated/Essential Fatty Acids

The term "essential fatty acids" refers to those polyunsaturated fatty acids that must be provided by foods because these cannot be synthesized in the body yet are necessary for health. Essential fatty acids are important constituents of all cell membranes and alter membrane fluidity and thus, determine and influence the behavior of membrane-bound enzymes and receptors. Essential fatty acids are lipids that are not metabolized by the body but are involved in various biological processes, and produce many compounds when they are metabolized (Kapalka, 2009). Essential fatty acids play an important role in the life and death of cardiac cells, immune system function and blood pressure regulation. Essential fatty acids must be obtained from food. They fall into two categories—omega-3 and omega-6.

Omega-3 fatty acids have in common a final carbon–carbon double bond in the ω-3 position (i.e., the third bond from the methyl end of the fatty acid), whereas ω-6 fatty acids have it in the ω-6 position (i.e., the sixth bond from the methyl end of the fatty acid) (Das, 2006). The importance of ω-3 and ω-6 designation is that n-end is never changed during physiological transformation in the human body as it is the most stable energetically. The double bonds in these ω-3 fatty acids are in the *cis*-configuration i.e., the two hydrogen atoms are on the same side of the double bond. α-linolenic acid (ALA), eicosapentaenoic acid (EPA) and docosahexaenoic acid (DHA) have various properties for which they can be classified as functional foods. DHA is an ω-3 essential fatty acid shown to play important roles in synaptic transmission in the brain during fetal development.

Omega-3 and omega-6 fatty acids are precursors to important compounds called eicosanoids. Eicosanoids are powerful hormones that control many other hormones and important body functions, such as the central nervous system and the immune system. Eicosanoids derived from omega-6 fatty acids are known to increase blood pressure, immune response and inflammation. In contrast, eicosanoids derived from omega-3 fatty acids are known to have heart-healthy effects. Given the contrasting effects of the omega-3 and omega-6 fatty acids, a proper dietary balance between the two must be achieved to ensure optimal health benefits.

1.5 OMEGA-3 FATTY ACIDS

Important omega-3 FAs include linolenic acid (18:3n-3, ALA), stearidonic acid (18:4n-3, SDA), eicosapentaenoic acid (C20:5n-3, EPA) and docosahexaenoic acid (C22:6n-3, DHA). Delta-6-desaturase is the key enzyme which controls the conversion of ALA to SDA. As the human body has inadequate functioning of this enzyme, dietary supply of SDA, EPA and DHA is a must (Stark et al., 2008).

1.5.1 Alpha-Linolenic Acid (18:3n-3, ALA)

Alpha-linolenic acid (linolenic acid, ALA) is a plant-based ω-3 PUFA mainly found in flax seed (Patenaude et al., 2009), chia (Vuksan et al., 2007) and perilla (Kurowska et al., 2003). HUFA omega-3 FAs (having $\leq$20 carbon atoms and $\leq$3 C-C double bonds) and other long-chain metabolites with significant potential benefits can be synthesized by ALA. Approximately 6–12% of ALA is converted to EPA and 3.8–9% to DHA, with reproductive-age women having high conversion rates while adult men on a high saturated fat diet have lower conversion (Gerster, 1998; Burdge and Wootton, 2002).

Other than its conversion to HUFAs, ALA has several physiological functions including cardio-protective effects, modulation of inflammatory response and positive effect on the functioning and behavior of the central nervous system (Stark et al., 2008). Due to many potential benefits in and of itself, dietary supply of ALA is essential even if other HUFA (SDA, EPA and DHA) supplements are taken (Stark et al., 2008).

1.5.2 Stearidonic Acid (18:4n-3, SDA)

Stearidonic acid (SDA) is another ω-3 LC-PUFA, naturally present in small amounts in fish and other seafood. Compared to other HUFAs such as EPA and DHA (15–20%), SDA contributes only 0.5–2% of total FAs. Mackerel is a rich source, with approximately 7% SDA, while echium oil from *Echium plantagineum* (approximately 12.5%) is an excellent vegetarian source of SDA (Chilton et al., 2008). Compared to ALA, SDA has been proved to be much more effective precursor of EPA in human studies (Miles et al., 2004; Horia and Watkins, 2005). It was found that the SDA was 3–5 times more effective than ALA in increasing tissue EPA content in erythrocyte and plasma phospholipids (James et al., 2003; Harris et al., 2008). Apart from being the precursor to EPA, SDA has many beneficial biological effects including hypotriglyceridemic, anti-inflammatory properties (Surette et al., 2004; Chilton et al., 2008; Whelan, 2009). Additionally it competes with endogenous arachidonic acid (AA) metabolism and also interferes in 2-series prostaglandin biosynthesis from AA, resulting in lowered inflammation, weak inhibition of platelet aggregation and reduced prostaglandin-dependent tumorigenesis (Kockmann et al., 1989; Petrik et al., 2000; Kelavkar et al., 2009).

1.5.3 Eicosapentaenoic Acid (20:5n-3, EPA)

Eicosapentaenoic acid (EPA) is an ω-3 HUFA, either biosynthesized from plant dietary PUFA or obtained directly from fish (salmon, herring, mackerel, etc.) and fish oils (cod liver oil). About 6–12% of ALA is converted to EPA (Gerster, 1998; Burdge and Wootton, 2002), which has strong anti-inflammatory actions. In contrast to arachidonic acid, which produces high levels of pro-inflammatory 2-series prostaglandins and 4-series leukotrienes, high EPA level results in production of anti-inflammatory 3-series prostaglandins. Intake of a diet rich in EPA-rich fish oil results in reduced pro-inflammatory cytokines (Kim et al., 2009; Kelavkar et al., 2009) whereas it is beneficial in rheumatoid arthritis and asthma (Miles et al., 2004). Increased dietary EPA intake also resulted in reduced low-density cholesterol, low-plasma triglycerides and reduced major as well as nonfatal coronary events (Yokoyama et al., 2007) while resulting in increased levels of high-density lipoproteins (Holub, 1988).

1.5.4 Docosahexaenoic Acid (22:6n-3, DHA)

Docosahexaenoic acid (DHA) is another important ω-3 HUFA which may either be biologically synthesized from EPA or directly obtained from oily fish including salmon, herring, mackerel, etc. and their oils. In general, only 4–9% of total ALA is bio-converted to DHA (Gerster, 1998; Burdge and Wootton, 2002) which can improve the functioning of cell membranes as well as cellular signaling. Comparatively, DHA is better than EPA in improving the membrane fluidity of endothelial cells (Hashimoto et al., 1999), which may be attributed to its ability to influence the functioning of membrane proteins, enzymes and ion channels (Hirafuji et al., 2003), resulting in improved cellular signaling. Due to this, DHA has been found to have unique beneficial effects on the heart and to be effective in preventing/curing senile dementia, depression and certain visual dysfunctions and other conditions (Hirafuji et al., 2003). Compared to EPA, DHA was found to have more promising effects on blood pressure and heart rate, especially in hyperlipidemic men (Kurowska et al., 2003; Landmark and Alm, 2006; Takeuchi et al., 2007; Ogawa et al., 2009). This can be attributed to the relatively low ability of men to convert ALA to DHA (Kitson et al., 2010). However, DHA was found to be a more effective anti-thrombotic agent in comparison to EPA (Woodman et al., 2003a). DHA was also found to reduce glycation and oxidation of LDL by increasing its particle size in type 2 diabetic patients (Woodman et al., 2003b).

1.6 OMEGA-6 FATTY ACIDS

The major ω-6 PUFAs are linoleic acid (C18:2n-6, LA) and arachidonic acid (C20:4n-6, AA). Apart from these, other important n-6 EFAs are gamma (γ)-linolenic acid (C18:3n-6, GLA) and dihomo-gamma-linolenic acid (C20:3n-6, DGLA). Omega-6 fatty acids are predominantly found in vegetable oils such as sunflower oil, safflower oil, sesame oil and corn oil, and they are also present in vegetables, cereals, legumes and tubers as invisible fat (Singh, 2005). In contrast to omega-3 EFAs, omega-6 EFAs are pro-inflammatory agents which can promote aggregation of platelets and trigger glutamate and oxygen free radical production in the brain (Singh, 2005).

1.6.1 Linoleic Acid (18:2n-6, LA)

Linoleic acid (LA) is the parent omega-6 fatty acid, which absolutely cannot be synthesized in the human body and must be supplied by diet or supplementation. Of total dietary ω-6 PUFAs, about 85–90% is LA (Franzen-Castle and Ritter-Gooder, 2010), which is commonly found in oil seeds, meat and grains. Sunflower, safflower and corn are richest; sesame, soybean and almond are medium; olive, peanut and canola oils are low; while coconut and palm kernel are the poorest sources of LA. It is converted into other fatty acids such as GLA and AA in the human body as and when needed. Studies reported lowering of blood cholesterol by replacing dietary SFAs with omega-6 EFAs (Mensink and Katan, 1992; Mensink and World Health Organization, 2016; Hooper et al., 2018) but convincing evidence of clinical cardiovascular benefits have not found (Ramsden et al., 2013; Hamley, 2017; Hooper et al., 2018). LA might reduce the blood cholesterol by (1) upregulation of low-density lipoprotein (LDL) receptors and redistribution of LDL cholesterol from plasma to tissue, (2) increased production of bile acid and cholesterol catabolism and (3) decreased conversion of very-low-density lipoprotein (VLDL) to LDL (Fernandez and West, 2005).

1.6.2 Arachidonic Acid (20:4n-6, AA)

Arachidonic acid (AA), a 20-carbon PUFA with 4 *cis*-double bonds, is found ubiquitously in plasma membrane bounded with phospholipids. In animals, but not plants, LA can be converted to AA at a limited rate. Therefore, AA is absent from oils and fats of plant origin but found in lesser amounts in meat, poultry and eggs. It is reported to the be most potent pro-inflammatory agent among n-6 PUFAs (Singh, 2005).

1.6.3 Gamma-Linolenic Acid (18:3n-6, GLA)

Gamma-linolenic acid (GLA) is a conditionally essential FA; can be produced in the body from LA, but not in sufficient quantity. Therefore, it can be obtained from dietary sources such as evening primrose oil (Hoy et al., 1983), black currant oil and borage oil (Lawson and Hughes, 1988). Organ meat and human milk are also poor sources of GLA.

1.6.4 Dihomo-Gamma-Linolenic Acid (20:3n-6, DGLA)

Eicosatrienoic acid, also known as dihomo-gamma-linolenic acid (DGLA), is the elongation product of GLA. It is an important constituent of neuronal membrane (Brownlee et al., 2016) and can be converted to prostaglandin E1, which can inhibit platelet aggregation (Ha and Bhagavan, 2011). By competing with AA for pathway enzymes, it suppresses the AA's eicosanoid production. Nowadays, DGLA is commercially produced by using *Mortierella* fungi (Sakuradani et al., 2010).

1.7 OMEGA-9 FATTY ACIDS

Oleic acid (18:1n-9) and nervonic acid (24:1n-9) are two major n-9 fatty acids. These fatty acids act synergistically with other biomolecules and PUFAs to contribute to metabolism and the maintenance of overall health (Dhopeshwarkar and Mead, 1961; Lowry and Tinsley, 1966; Maskrey et al., 2011).

1.8 CONCLUSION

The most important fats are those that the body cannot make and thus must come from the food we eat. Essential fatty acids have significant effects on overall health and well being. Essential fatty acids need more research, as they can offer potential solutions for diseases and there is a need to optimize the concentration of fatty acids required to show therapeutic benefits. The different and numerous ratios of omega-3 to omega-6 fatty acids need to be researched further to obtain a uniform value for their use for various health benefits.

REFERENCES

Aron, H., and Uber den Nahwert. 1919. (On the nutritional value). *Biochemische Zeitschrift* 92: 211–233.

Brownlee, M., Aiello, L. P., Cooper, M. E., Vinik, A. I., Plutzky, J., and Boulton, A. J. M. 2016. Chapter 33: Complications of diabetes mellitus. In: Melmed, S., Polonsky, K. S., Larsen, P. R., and Kronenberg, H.M. (eds.), *Williams Textbook of Endocrinology*, 13th ed. Philadelphia, PA: Elsevier, vol. 1484, p. 581.

Buist, P. H. 2010. Unsaturated fatty acids. Comprehensive natural products II. *Chemistry and Biology* 1: 5–33.

Burdge, G. C., and Wootton, S. A. 2002. Conversion of α-linolenic acid to eicosapentaenoic, docosapentaenoic and docosahexaenoic acids in young women. *The British Journal of Nutrition* 88(4): 411–420.

Burr, G. O., and Burr, M. M. 1929. A new deficiency disease produced by the rigid exclusion of fat from the diet. *The Journal of Biological Chemistry* 82: 345–367.

Burr, G. O., and Burr, M. M. 1930. On the nature and role of the fatty acids essential in nutrition. *The Journal of Biological Chemistry* 86: 587–621.

Burr, G. O., Brown, J. B., Kass, J. P., and Lundberg, W. O. 1940. Comparative curative values of unsaturated fatty acids in fat deficiency. *Proceedings of the Society for Experimental Biology and Medicine* 44(1): 242–245.

Chilton, F. H., Rudel, L. L., Parks, J. S., Arm, J. P., and Seeds, M. C. 2008. Mechanisms by which botanical lipids affect inflammatory disorders. *The American Journal of Clinical Nutrition* 87(2): 498S–503S.

Das, U. N. 2006. Essential fatty acids-a review. *Current Pharmaceutical Biotechnology* 7(6): 467–482.

Dhopeshwarkar, G. A., and Mead, J. F. 1961. Role of oleic acid in the metabolism of essential fatty acids. *Journal of the American Oil Chemists Society* 38(6): 297–301.

Fernandez, M. L., and West, K. L. 2005. Mechanisms by which dietary fatty acids modulate plasma lipids. *The Journal of Nutrition* 135(9): 2075–2078.

Food and Nutrition Board, Institute of Medicine. 2002. *Dietary Fats: Total Fat and Fatty Acids. Dietary Reference Intakes for Energy, Carbohydrate, Fiber, Fat, Fatty Acids, Cholesterol, Protein, and Amino Acids.* Washington, DC: National Academies Press, pp. 422–541.

Gerster, H. 1998. Can adults adequately convert a-linolenic acid (18: 3n-3) to eicosapentaenoic acid (20: 5n-3) and docosahexaenoic acid (22: 6n-3)? *International Journal for Vitamin and Nutrition Research* 68(3): 159–173.

Ha, C. E., and Bhagavan, N. V. 2011. *Essentials of Medical Biochemistry: with Clinical Cases.* San Diego, CA: Academic Press.

Hamley, S. 2017. The effect of replacing saturated fat with mostly n-6 polyunsaturated fat on coronary heart disease: A meta-analysis of randomised controlled trials. *Nutrition Journal* 16(1): 30.

Harris, W. S., Lemke, S. L., Hansen, S. N. et al. 2008. Stearidonic acid-enriched soybean oil increased the omega-3 index, an emerging cardiovascular risk marker. *Lipids* 43(9): 805–811.

Hashimoto, M., Hossain, M. S., Yamasaki, H., Yazawa, K., and Masumura, S. 1999. Effects of eicosapentaenoic acid and docosahexaenoic acid on plasma membrane fluidity of aortic endothelial cells. *Lipids* 34(12): 1297–1304.

Hirafuji, M., Machida, T., Hamaue, N., and Minami, M. 2003. Cardiovascular protective effects of n-3 polyunsaturated fatty acids with special emphasis on docosahexaenoic acid. *Journal of Pharmacological Sciences* 92(4): 308–316.

Holub, B. J. 1988. Dietary fish oils containing eicosapentaenoic acid and the prevention of atherosclerosis and thrombosis. *CMAJ: Canadian Medical Association Journal* 139(5): 377.

Hoy, C.E., Holmer, G., Kaur, N., Byrjalsen, I., and Kirstein, D. 1983. Acyl group distribution tissue lipids of rats fed evening primrose oil (γ-linolenic plus linoleic acid) or soybean oil (α-linolenic plus linoleic acid). *Lipids* 18: 760–771.

Hooper, L., Al-Khudairy, L., Abdelhamid, A. S., et al. 2018. Omega-6 fats for the primary and secondary prevention of cardiovascular disease. *Cochrane Database of Systematic Review* 7.

Horia, E., and Watkins, B. A. 2005. Comparison of stearidonic acid and α-linolenic acid on PGE2 production and COX-2 protein levels in MDA-MB-231 breast cancer cell cultures. *The Journal of Nutritional Biochemistry* 16(3): 184–192.

Jakobsen, M. U., O'Reilly, E. J., Heitmann, B. L. et al. 2009. Major types of dietary fat and risk of coronary heart disease: A pooled analysis of 11 cohort studies. *The American Journal of Clinical Nutrition* 89(5): 1425–1432.

James, M. J., Ursin, V. M., and Cleland, L. G. 2003. Metabolism of stearidonic acid in human subjects: Comparison with the metabolism of other n-3 fatty acids. *The American Journal of Clinical Nutrition* 77(5): 1140–1145.

Kapalka, G. M. 2009. *Nutritional and Herbal Therapies for Children and Adolescents: A Handbook for Mental Health Clinicians*. London: Academic Press.
Kelavkar, U. P., Hutzley, J., McHugh, K., Allen, K. G., and Parwani, A. 2009. Prostate tumor growth can be modulated by dietarily targeting the 15-lipoxygenase-1 and cyclooxygenase-2 enzymes. *Neoplasia (New York, NY)* 11(7): 692.
Kim, J. Y., Park, H. D., Park, E., Chon, J. W., and Park, Y. K. 2009. Growth-inhibitory and proapoptotic effects of alpha-linolenic acid on estrogen-positive breast cancer cells. *Annals of the New York Academy of Sciences* 1171(1): 190.
Kitson, A. P., Stroud, C. K., and Stark, K. D. 2010. Elevated production of docosahexaenoic acid in females: Potential molecular mechanisms. *Lipids* 45(3): 209–224.
Kockmann, V., Spielmann, D., Traitler, H., and Lagarde, M. 1989. Inhibitory effect of stearidonic acid (18: 4 n–3) on platelet aggregation and arachidonate oxygenation. *Lipids* 24(12): 1004–1007.
Kurowska, E. M., Dresser, G. K., Deutsch, L., Vachon, D., and Khalil, W. 2003. Bioavailability of omega-3 essential fatty acids from perilla seed oil. *Prostaglandins, Leukotrienes and Essential Fatty Acids* 68(3): 207–212.
Landmark, K., and Alm, C. S. 2006. Alpha-linolenic acid, cardiovascular disease and sudden death. *Tidsskrift for den Norske Laegeforening: Tidsskrift for Praktiskmedicin, Nyraekke* 126(21): 2792–2794.
Lawson, L. D. and Hughes, B.G. 1988. Absorption of eicosapentaenoic acid and docosahexaenoic acid from fish oil triacylglycerols or fish oil ethyl esters co-ingested with a high-fat meal. *Biochemical and Biophysical Research Communications* 156(2): 960–963.
Lowry, R. R., and Tinsley, I. J. 1966. Oleic and linoleic acid interaction in polyunsaturated fatty acid metabolism in the rat. *The Journal of Nutrition* 88(1): 26–32.
Maskrey, B. H., Megson, I. L., Whitfield, P. D., and Rossi, A. G. 2011. Mechanisms of resolution of inflammation: A focus on cardiovascular disease. *Arteriosclerosis, Thrombosis, and Vascular Biology* 31(5): 1001–1006.
Mensink, R. P., and Katan, M. B. 1992. Effect of dietary fatty acids on serum lipids and lipoproteins. A meta-analysis of 27 trials. *Arteriosclerosis and Thrombosis: A Journal of Vascular Biology* 12(8): 911–919.
Mensink, R. P., and World Health Organization 2016. Effects of saturated fatty acids on serum lipids and lipoproteins: A systematic review and regression analysis. WHO, pp. 1–72.
Miles, E. A., Banerjee, T., and Calder, P. C. 2004. The influence of different combinations of γ-linolenic, stearidonic and eicosapentaenoic acids on the fatty acid composition of blood lipids and mononuclear cells in human volunteers. *Prostaglandins, Leukotrienes and Essential Fatty Acids* 70(6): 529–538.
Ogawa, A., Suzuki, Y., Aoyama, T., and Takeuchi, H. 2009. Dietary alpha-linolenic acid inhibits angiotensin-converting enzyme activity and mRNA expression levels in the aorta of spontaneously hypertensive rats. *Journal of Oleo Science* 58(7): 355–360.

Patenaude, A., Rodriguez-Leyva, D., Edel, A. L., et al. 2009. Bioavailability of α-linolenic acid from flaxseed diets as a function of the age of the subject. *European Journal of Clinical Nutrition* 63(9): 1123–1129.

Petrik, M. B. H., McEntee, M. F., Johnson, B. T., Obukowicz, M. G., and Whelan, J. 2000. Highly unsaturated (n-3) fatty acids, but not α-linolenic, conjugated linoleic or γ-linolenic acids, reduce tumorigenesis in Apc Min/+ mice. *The Journal of Nutrition* 130(10): 2434–2443.

Punia, S., Sandhu, K. S., Siroha, A. K., and Dhull, S. B. 2019. Omega 3-metabolism, absorption, bioavailability and health benefits: A review. *PharmaNutrition* 10: 100162.

Quackenbush, F. W., Kummerow, F. A., and Steenbock, H. 1942. The effectiveness of linoleic, arachidonic, and linolenic acids in reproduction and lactation. *The Journal of Nutrition* 24(3): 213–224.

Ramsden, C. E., Zamora, D., Leelarthaepin, B. et al. 2013. Use of dietary linoleic acid for secondary prevention of coronary heart disease and death: Evaluation of recovered data from the Sydney Diet Heart Study and updated meta-analysis. *British Medical Journal* 346: e8707.

Rustan, A. C., and Drevon, C. A. 2005. Fatty acids: Structures and properties. *Encyclopedia of Life SciencesS*. Chichester: Wiley.

Sakuradani, E., Ando, A., Ogawa, J., and Shimizu, S. 2010. Arachidonic acid-producing Mortierellaalpina: Creation of mutants, isolation of the related enzyme genes, and molecular breeding. In: Cohen, Z., and Ratledge, C. (eds.), *Single Cell Oils*. AOCS Press, pp. 29–49.

Singh, M. 2005. Essential fatty acids, DHA and human brain. *The Indian Journal of Pediatrics* 72(3): 239–242.

Stark, A. H., Crawford, M. A., and Reifen, R. 2008. Update on alpha-linolenic acid. *Nutrition Reviews* 66(6): 326–332.

Surette, M. E., Edens, M., Chilton, F. H., and Tramposch, K. M. 2004. Dietary echium oil increases plasma and neutrophil long-chain (n-3) fatty acids and lowers serum triacylglycerols in hypertriglyceridemic humans. *The Journal of Nutrition* 134(6): 1406–1411.

Takeuchi, H., Sakurai, C., Noda, R. et al. 2007. Antihypertensive effect and safety of dietary α-linolenic acid in subjects with high-normal blood pressure and mild hypertension. *Journal of Oleo Science* 56(7): 347–360.

Turpeinen, O. 1937. Further studies on the unsaturated fatty acids essential in nutrition. *Journal of Nutrition* 15(4): 351–366.

Vuksan, V., Whitham, D., Sievenpiper, J. L., et al. 2007. Supplementation of conventional therapy with the novel grain Salba (*Salvia hispanica* L.) improves major and emerging cardiovascular risk factors in type 2 diabetes: Results of a randomized controlled trial. *Diabetes Care* 30(11): 2804–2810.

Whelan, J. 2009. Dietary stearidonic acid is a long chain (n-3) polyunsaturated fatty acid with potential health benefits. *The Journal of Nutrition* 139(1): 5–10.

Whitney, E. N., Cataldo, C. B., and Rolfes, S. R. 1998. *Understanding Normal and Clinical Nutrition*, 5th ed. Belmont, CA: West/Wadsworth, pp. 141–175.

Woodman, R. J., Mori, T. A., Burke, V., et al. 2003a. Effects of purified eicosapentaenoic acid and docosahexaenoic acid on platelet, fibrinolytic and vascular function in hypertensive type 2 diabetic patients. *Atherosclerosis* 166(1): 85–93.

Woodman, R. J., Mori, T. A., Burke, V., et al. 2003b. Docosahexaenoic acid but not eicosapentaenoic acid increases LDL particle size in treated hypertensive type 2 diabetic patients. *Diabetes Care* 26(1): 253–253.

Yokoyama, M., Origasa, H., Matsuzaki, M. et al. 2007. Effects of eicosapentaenoic acid on major coronary events in hypercholesterolaemic patients (JELIS): A randomised open-label, blinded endpoint analysis. *The Lancet* 369(9567): 1090–1098.

2

Sources

Plants, Animals and Microbial

Sanju Bala Dhull and Sneh Punia

Contents

2.1 INTRODUCTION

Most fats needed, including cholesterol, saturated fatty acids (SFAs) and unsaturated fatty acids (UFAs), can be synthesized by the human body except two families of fatty acids (FAs), i.e., the ω-6 (n-6) and the ω-3 (n-3) fatty acids, which are considered essential fatty acids (EFAs). These are termed as essential because the parent ω-6 fatty acid, linoleic acid (C18:2n-6, LA), and the parent ω-3 fatty acid, α-linolenic acid (C18:3n-3, ALA), which are absolutely not biosynthesized in the human body and must be supplied by diet or supplementation (Table 2.1).

Arachidonic acid (C20:4n-6, AA), another ω-6 FA, can be synthesized from LA in the human body, while ALA can be converted to other ω-3 FAs including eicosapentaenoic acid (C20:5n-3, EPA), docosapentaenoic acid (C22:5n-3, DPA) and docosahexaenoic acid (C22:6n-3, DHA). However, the rate of bioconversion of ALA in EPA, DPA and DHA is very low (Domenichiello et al. 2015; Burdge and Calder 2005) and insufficient; therefore, these n-3 FAs are considered EFAs too.

Both omega-3 (ω-3) and omega-6 (ω-6) polyunsaturated fatty acids (PUFAs) are entirely obtained from the diet; but as suggested by nutrition experts, an ω 6:ω 3 ratio of 5:1 or less is desirable (WHO/FAO 1994). Some powerful hormones known as eicosanoids (Figure 2.1) are derived

Table 2.1 Possible Sources of Some Major Physiologically Occurring Fatty Acids in Human Body

Fatty Acids	Sources (Dietary and/or Endogenous)
Palmitic acid (16:0)	Dietary and endogenous
Stearic acid (18:0)	Dietary and endogenous (from palmitic acid)
Oleic acid (cis 18:1)	Dietary and endogenous (from stearic acid)
Oleic acid (trans 18:1)	Dietary
Linoleic acid (18:2n-6)	Dietary
α-linolenic acid (18:3n-3)	Dietary
Arachidonic acid (20:4 n-6)	Dietary & endogenous (from linoleic acid)
Eicosapentaenoic acid (20:5n-3) and Docosahexaenoic acid (22:6n-3)	Dietary (needs arachidonic acid to be present) Limited conversion from arachidonic acid to eicosapentaenoic acid & docosahexaenoic acid

(*Source:* adapted from Kaur, Chugh and Gupta 2014)

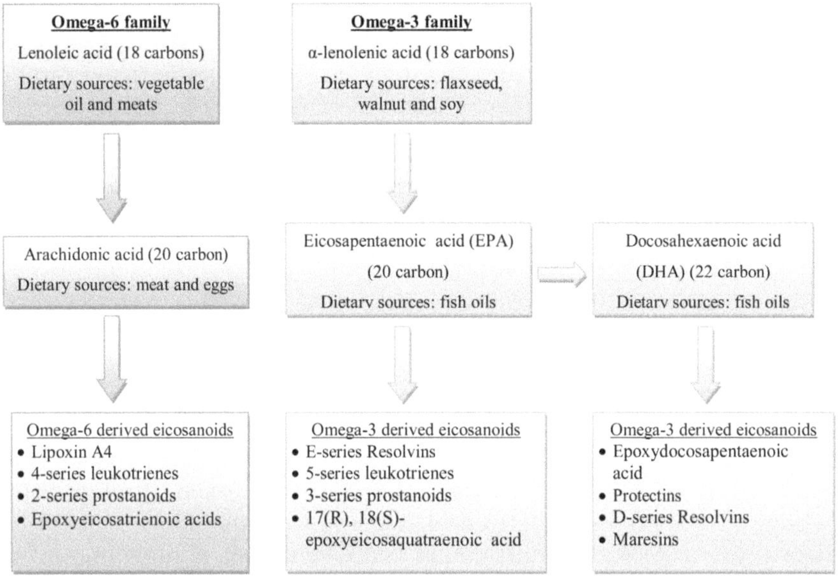

Figure 2.1 Essential fatty acid family

from ω-3 and ω-6 fatty acids and control several other hormones as well as regulate important body functions including the central nervous system and the immune system. ω-6 FA-derived eicosanoids are known to be proinflammatory and prothrombotic and increase blood pressure and immune response, while ω-3 FA-derived eicosanoids are reported to have healthy effects on heart, to be anti-inflammatory and to prevent blood clotting. Since ω-3 and ω-6 fatty acids are known to have contrasting effects, a proper balance of dietary intake between the two must be achieved for optimal health benefits.

Nowadays, modern diets are mainly characterized by a high proportion of fast foods, snacks, meats and oils from seeds having high amounts of SFAs but low PUFA content (Fernández-San Juan 2000). Comparison of fast food composition with foods rich in fish such as typical Japanese food (Kamei et al. 2002) or Mediterranean food (Ambring et al. 2006) showed that people consuming Japanese or Mediterranean food had ω-6:ω-3 ratios close to 2:1 in blood, while the values reached to much higher levels (25:1) than desired in people consuming fast foods. Therefore, it is advised by nutritionists to consume diets having more fish and green vegetables to prevent cardiovascular and other diseases. Generally, it is very difficult to change the nutritional preferences of a whole society. But products such as nutritional supplements or functional foods enriched with EFAs can be developed to enrich the diet and a good ω-6:ω-3 ratio in blood can be reached without altering dietary choices too much.

2.2 STATUS OF ω-6 FATTY ACIDS IN DIET

Linoleic acid (LA) is present abundantly in the diets of most of the people. During the past century, the intake of LA has doubled due to increased use of oils from safflower, corn and cotton. As mentioned earlier, a balanced ω-6:ω-3 ratio is desirable; therefore, diets rich in LA or poor in ω-3 might result in hypertension, chronic inflammation and cardiovascular problems. During metabolism, ω-6 and ω-3 fatty acids compete for the same enzymes (*delta-5 and delta-6 desaturase*) for their bioconversion into specific eicosanoids (Martins et al. 2008). For this reason, through competitive inhibition which fatty acid is going to dominate is decided by dietary LA or ALA (Arterburn, Hall and Oken 2006). A high amount of LA is known to inhibit ALA metabolism to EPA and DHA, which might further decrease with increasing age (Simopoulos 2002a). The fatty acid composition of different oils is given in Table 2.2.

2.3 STATUS OF ω-3 FATTY ACIDS IN DIET

Parent α-linolenic acid (ALA) has three carbon–carbon double bonds and is very sensitive to oxygen, light and temperature. It is destructed easily; five times faster than LA, and becomes toxic if not protected properly. In contrast to LA, currently the average consumption of ω-3 fatty acids has been reduced to less than 20% compared to diets 150 years ago. Generally most of the population (about 95–99%) do not get sufficient ω-FAs required for daily needs and good health, making it most essential and therapeutic among rest of fatty acids and essential nutrients (20 minerals, 14 vitamins, 8–11 amino acids) (Kaur, Chugh and Gupta 2014). Although increased consumption of meat slightly increased dietary level of AA (about 0.2–1.0 mg/day) (Phinney et al. 1990), ALA (2.92 g/day) (Raper, Cronin and Exler 1992) and EPA and DHA (48 and 72 mg/day, respectively) (Kaur et al. 2014); but a relative and absolute decrease in the amount of ω-3 FAs is observed, which has led to an imbalance and increase in the ratio of ω-6:ω-3.

Table 2.2 Average Linoleic and α-Linolenic Acid Content (%) of Different Oils

Sr. No.	Oils	Linoleic Acid 18:2 (n-6) (%)	α-Linolenic Acid 18:3 (n-3) (%)	Total Unsaturated Fatty Acids (%)
1	Safflower	73.0	0.5	86.3
2	Corn	57.3	0.8	82.8
3	Sunflower	66.4	0.3	88.5
4	Cotton seed	50.3	0.4	69.6
5	Sesame	40.0	0.5	80.5
6	Soybean	50.8	6.8	80.7
7	Olive	8.2	0.7	81.4
8	Peanut	31.0	1.2	77.8
9	Coconut	1.8	–	7.9
10	Cocoa butter	2.8	0.2	36.0
11	Rapeseed (zero erucic acid)	22.2	11.0	88.0
12	Rapeseed (high erucic acid)	12.8	8.6	88.4
13	Palm kernel	1.5	–	7.9
14	Palm	9.0	0.3	47.7
15	Cashew	17.0	0.4	73.8
16	Almond	18.2	0.5	87.7
17	Walnut	61.0	6.7	86.5
18	Chestnut	35.0	4.0	75.5
19	Butter	2.3	1.4	32.6

(*Source:* adapted from Kaur, Chugh and Gupta 2014)

2.4 SOURCES OF ESSENTIAL FATTY ACIDS

Naturally, ALA, a parent EFA, is a primary dietary fatty acid found mainly in seeds and seed oil which is bioconverted to EPA and DHA in the human body. Fish and other seafood are rich sources of ω-3 FAs (DHA and EPA), while ω-6 FAs are mainly found in vegetable oils such as sunflower, corn and soybean (Martins et al. 2008).

2.4.1 Omega-3 Fatty Acids

Omega-3 (ω-3, w-3, n-3) FAs including ALA, EPA and DHA are known as the common and most important long-chain PUFAs, having several double bonds started at the position 3 from the terminal methyl ends. Two primary ω-3 PUFAs (i.e., EPA and DHA), referred as the "miracle food of 21st century" (Swinbanks 1993; Bourre 2007; Amiri-Jami and Griffiths 2010), serve as bioactive lipids (Amjad Khan et al. 2017) and have shown multiple health benefits proving to be very essential for human health (Islam et al. 2018; Tocher et al. 2019; Punia et al. 2019). As aforementioned, the yield of bioconversion of ALA to EPA and finally to DHA from DPA is very low in the human body and insufficient (Domenichiello et al. 2015; Burdge and Calder 2005) and it is essential to maintain a sufficient dietary supply of ω-3 fatty acids. Different ω-3 fatty acids can be obtained from various sources (Figure 2.2) including the following:

2.4.1.1 Plant Sources

The primary dietary sources of the parent ω-3 FA (i.e., ALA) are seeds and their oils including flaxseeds, perilla seeds and chia seeds (Kris-Etherton et al. 2000; Racine and Deckelbaum 2007). Flaxseed oil, also known as linseed oil, which is extracted from flaxseeds (*Linium usitatissimum*), is the richest source available for ALA, containing up to 50% of it when cold pressed (Breivik 2007). Echium oil extracted from seeds of the *Echium plantagineum* plant contains up to 33% ALA and 12.5% stearidonic acid (18:4n-3, SDA) (Breivik 2007). SDA, another omega-3 fatty acid, follows ALA in the n-3 metabolic pathway (Bailey 2009). Walnut oil also found rich in ALA, containing up to 10.4% of it (Simopoulos 2002b). Other plant sources include different nuts, canola, soybean, dark green leafy vegetables and some fruits (Tables 2.1 and 2.2).

2.4.1.2 Animal and Marine Sources

Among animal sources, ω-3 PUFAs are mainly found in marine organisms such as fish and other seafood, red and white meats, as well as egg yolk. In all animals, the cell membrane phospholipids are constituted of n-3 PUFAs reaching up to a level of 0.40 mg per 100 g of meat, but EPA and DHA are poorly present in the meat products. For dietary patterns having low fish intake, consumption of meat from poultry, pork, beef and lamb may add small amounts of DHA.

Marine organisms including fish, seafood and several algae are the most important natural sources of omega-3 FAs. Some marine animals

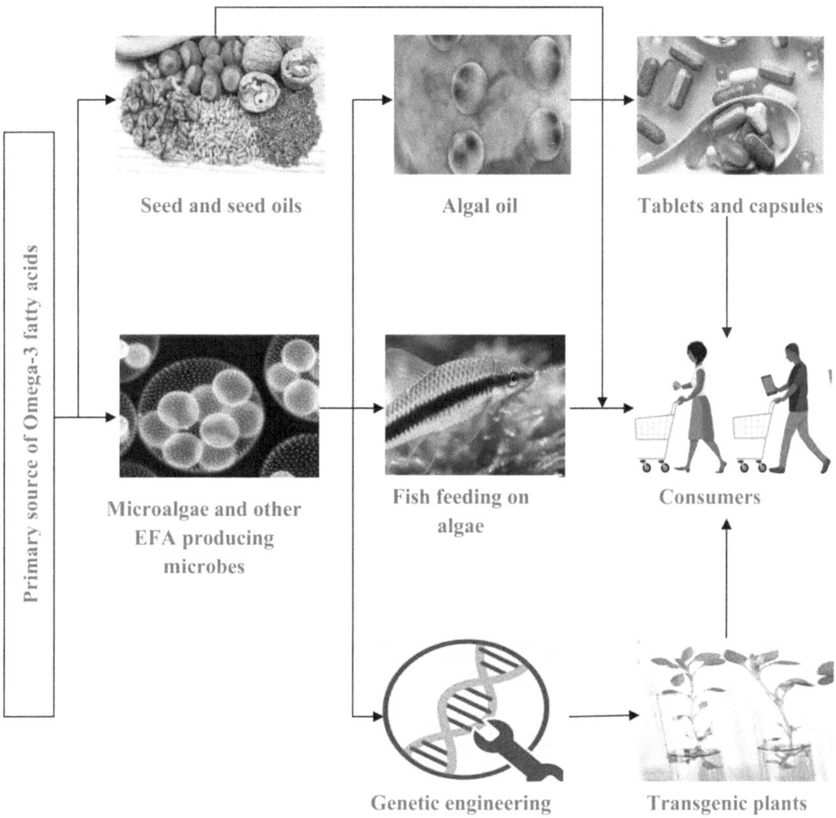

Figure 2.2 Sustainable sources of omega-3 polyunsaturated fatty acids

such as mackerel, herring and salmon are served as main sources of EPA and DHA (Kris-Etherton et al. 2000; Pacetti et al. 2013). These organisms are directly or indirectly fed from marine phytoplankton, which primarily produces the ω-3 PUFAs in the trophic chain (Rubio-Rodríguez et al. 2010; Ryckebosch et al. 2012). Studies on marine lipidology concluded that marine oil PUFAs are generally present in the form of triglycerides, which can be further converted to different types of lipids (wax esters, phospholipids, etc.) (Bergé and Barnathan 2005). In addition, several fish oil studies about the positional distribution of FAs in triglycerides showed that PUFAs are mostly linked to the sn–2 position of the glycerol backbone (Ando et al. 1992), which makes it more stable towards oxidation reactions (Wijesundera et al. 2008).

Nowadays, fish and fish products, especially fish oils, are the richest sources of ω-3 fatty acids in the human diet. The highest levels of DHA and EPA were found in the food portions having some oily fishes such as *Scombridae*, *Clupeidae* and *Salmonidae* families (Table 2.3) (Mataix et al. 2003; Rubio-Rodríguez et al. 2010). However, some commonly consumed fish and seafood including cod, clams, shrimp, crabs and oysters are found to have low DHA content (Arbex et al. 2015). In addition, freshwater fish have been found with high FA content compared with marine fish, which was attributed to the high FA composition of freshwater phytoplankton (Pacetti et al. 2013). Despite this fact, phytoplankton of marine ecosystems have higher EPA and DHA content than freshwater phytoplankton (de Oliveira et al. 2016).

Fish oils are usually richer in ω-3 PUFAs in comparison with different seed oils (Table 2.4) (Mataix et al. 2003) and microalgae. Therefore, nowadays fish oils not only are used for ω-3 concentrate production but are useful ingredients themselves in the enrichment of different food products. EPA and DHA levels of different fish oils such as tuna, salmon and herring ranged between 862 and 1840 mg per 100 g of oil (Arbex et al. 2015).

Table 2.3 Omega-3 Fatty Acid (EPA and DHA) Content in Some Fish Species

Fish Species/Common Name	EPA (C20:5n-3) (%)	DHA (C22:6n-3) (%)
Scomber scombrus (mackerel)	1.10	2.56
Mullus surmuletus (red mullet)	0.91	1.66
Sardina pilchardus (sardine)	0.62	1.12
Salmo salar (salmon)	0.50	1.00
Thunnus thynnus (tuna)	0.24	0.98
Gadus morrhua (cod)	0.23	0.47
Luvarus imperialis (swordfish)	0.15	0.30
Conger conger (Conger eel)	0.15	0.43
Engraulis encrasicolus (anchovy)	0.14	0.80
Pagellus bogaraveo (sea bream)	0.12	0.61
Merluccius merluccius (hake)	0.10	0.54
Galeorhinus geleus (dogfish)	0.04	0.30

(*Source:* adapted from Mataix et al. 2003)

Although fish oil is nowadays used as a main source of n-3 PUFAs, it has some limitations of use because of its undesirable effects on heart, inflammatory bowel disease and cancer-causing implications (Verma et al. 2018). Also, rapidly declining numbers and species of fish, environmental toxins such as mercury and polychlorinated biphenyls, the unpleasant odor and taste of fish oils, their instability and the complicated purifying process are various other factors which necessitate some healthy, safe, sustainable alternative sources of PUFAs (Mahaffey et al. 2008; Bourdon et al. 2010; Lenihan-Geels et al. 2013).

2.4.1.3 Microbial Sources

Various microflora, which include oleaginous marine bacteria, algae and fungi, are some other sources of ω-3 PUFAs (Table 2.4) (Amiri-Jami and Griffiths 2010; Verma et al. 2020). Oil extraction from these microbial sources has several advantages compared to other sources (vegetable oils, fish oils) such as short life cycle, cheap and abundant raw material and easy large-scale production (Liang and Jiang 2013).

2.4.1.3.1 Bacteria

Most bacteria have a predominant and well-characterized pathway of fatty acid synthesis (Santos-Merino et al. 2018). Several marine gamma-proteobacteria and deep-sea bacteria produce ω-3 FAs such as EPA and DHA, which are predominantly present in membrane phospholipids (Allemann, Shulse and Allen 2019) and regulate membrane fluidity (Yazawa et al. 1988). In deep-sea bacteria, factors such as high hydrostatic pressure and low temperature change the fluid state of cell membranes to a non-fluid state (Hazel and Williams 1990). To counter such problematic changes, greater amounts of ω-3 PUFAs are incorporated into membrane phospholipids by these bacteria. Species of some oleaginous bacteria such as *Shewanella putrefaciens*, *Pneumatophorus japonicas* and *Alteromonas putrefaciens* were demonstrated as good alternative sources of EPA (Yazawa 1996; Yazawa et al. 1988a, 1988b). Studies also found that growth temperature inversely affected the cellular EPA production in *Shewanella* strain SCRC-2378 (Yazawa 1996).

By photosynthesis, cyanobacteria can capture sunlight energy and convert it into desirable products such as fatty acids. Also, genetically engineered cyanobacteria such as *Synechococcus elongatus* PCC 7942 with improved ω-3 PUFA production pathways can be useful in ω-3 fatty acid production (Santos-Merino et al. 2018). Mesophilic and pressure-sensitive

Table 2.4 Different Microbial Sources of Omega-3 Polyunsaturated Fatty Acids

Microorganism	Type	Fatty Acid Present	References
Shewanella putrefaciens	Bacteria	EPA	Yazawa (1996)
Pneumatophorus japonicus	Bacteria	EPA	Yazawa et al. (1988)
Alteromonas putrefaciens	Bacteria	EPA	Yazawa et al. (1988)
Photobacterium	Bacteria	EPA	Ryan, Farr, Visnovsky, Vyssotski and Visnovsky (2010)
Yarrowia lipolytica	Yeast	EPA	Xie et al. (2015)
Thraustochytrium aureum	Fungi	EPA, DHA	Ward and Singh (2005)
Mortierella	Fungi	EPA	Jareonkitmongkol et al. (1993); Jermsuntiea et al. (2011)
Pythium spp.	Fungi	EPA	Athalye et al. (2009)
Pythium irregulate	Fungi	EPA	Liang, Zhao, Strait and Wen (2012)
Trichoderma spp.	Fungi	EPA	Liang et al. (2012)
Aspergillus niger	Fungi	EPA, DHA	Gayathri et al. (2010)
Candida guilliermondii	Fungi	DHA, EPA	Gupta et al. (2012)
Thraustochytrium	Microalgae	EPA, DHA	Scott et al. (2011)
Nannochloropsis	Microalgae	EPA, DHA	Hu and Gao (2003); Patil et al. (2007); Pal et al. (2011); Van Wagenen et al. (2012)
Pinguiococcus pyrenoidosus	Microalgae	EPA, DHA	Sang et al. (2012)
Chlorella minutissima	Microalgae	EPA	Yongmanitchai and Ward (1991); Khozin-Goldberg et al. (2002)
Pavlova spp.	Microalgae	EPA, DHA	Hu et al. (2008); Carvalho et al. (2006); Guihéneuf et al. (2009)

(Continued)

Table 2.4 (Continued) Different Microbial Sources of Omega-3 Polyunsaturated Fatty Acids

Microorganism	Type	Fatty Acid Present	References
Schizochytrium sp.	Microalgae	DHA	Fan et al. (2007); Ethior et al. (2011)
Ulkenia	Microalgae	DHA	Doughman et al. (2007); Chen et al. (2007); Borowitzka (2013); Klok et al. (2014)
Cryptocodinium cohnii	Microalgae	DHA	Kyle (2001)
Dunaliella salina	Microalgae	EPA	Bhosale, Rajabhoj and Chaugule (2010)
Phaeodactylum tricornutum	Microalgae	EPA	Wen and Chen (2003); Ji et al. (2015)
Nannochloropsis sp.	Microalgae	EPA	Chaturvedi and Fujita (2006)
Nitzschia laevis	Microalgae	EPA	Xiao-Hong, Song-Yao, Wang and Mei-Fang (2007)
Porphyridium cruentum	Microalgae	EPA	Wen and Chen (2002)
Odontella aurita	Microalgae	EPA	Guihéneuf et al. (2009)
Cyclotella cryptica	Microalgae	EPA	Pahl et al. (2010)
Cylindrotheca sp.	Microalgae	EPA	Suman et al. (2012)
Isochrysis galbana	Microalgae	EPA, DHA	Guihéneuf et al. (2009)
Pavloba lutheria	Microalgae	EPA, DHA	Guihéneuf et al. (2009)
Phaeodactylum tricornutum	Microalgae	EPA, DHA	Hamilton et al. (2016)

(*Source:* adapted from Verma et al. 2020)

species of bacteria that are not adapted to high pressure and low temperature only produce scanty amounts of ω-3 PUFAs (Kato and Nogi 2001).

2.4.1.3.2 Fungi

A high percentage of EPA can be accumulated in the lipid fraction of some species of lower fungi (Ward and Singh 2005), which might be due to the diversity of their lipids and fatty acid biosynthetic pathways.

At low-temperature growing conditions, a large amount of EPA is produced in the mycelia of many filamentous fungi from genus *Mortierella* (Jareonkitmongkol et al. 1993; Jermsuntiea et al. 2011; Cordova and Alper 2018). Under normal medium and growth conditions, *Mortierella* sp. produce PUFAs such as γ-linolenic acid (GLA, C18:3) and AA, but under the stress condition, EPA and DHA are produced in their mycelial biomass (Vadivelan and Venkateswaran 2014). Another fungus, *Pythium acanthium*, was reported to accumulate high amounts of DHA when cultured in a media containing linseed oil (Salunke et al. 2014). On the other hand, regardless of culture type and time, *Thraustochytrium* sp. was found to synthesize a high amount of ω-3 PUFAs with 69–73% of DHA and 21–15% of docosapentaenoic acid (DPA, C22:5) (Furlan et al. 2017). This fungus was can accumulate 50% of its weight as lipids, being more than 25% of DHA. Among soil-borne fungi, *Tichoderma* sp. and *Aspergillus niger* were reported to produce significant amounts of LC-PUFAs (EPA and DHA) (Gayathri et al. 2010).

Many different species of mushrooms, either harvested from nature or cultivated under controlled conditions, are consumed worldwide (Table 2.5). Wild edible mushroom species vary from area to area, while *Agaricus, Lentinula* and *Pleurotus* genera are some of the most cultivated and commercialized genera in many countries (Sande et al. 2019b). Mushrooms are considered as healthy foods having well-balanced carbohydrate and protein content but with low fat concentrations (0.1–16.3%). Although low in lipid content, mushrooms contain PUFAs (linoleic, oleic and linolenic acids) as major constituents of their lipid profiles. Generally, linolenic, linoleic and oleic acid content in mushrooms ranges from 0.0 to 28.8%, from 0.0 to 81.1% and from 1.0 to 60.3%, respectively (Sande et al. 2019). Therefore, for vegetarian and vegan diets, mushrooms can be a desirable cholesterol-free but protein-, fiber- and PUFA-rich substitute for meat products. Moreover, mushrooms require relatively little area and water for growth, are produced easily and have a rich bioactive profile (Valverde, Hernández-Pérez and Paredes-López 2015). Mushroom lipids are a balance of unsaturated to saturated fatty acids; mainly consisting of MUFAs (monounsaturated fatty acids) and PUFAs, which classify them as a source of healthy lipids (Abugri, McElhenney and Willian 2016). *Basidiomycota* species mostly contains palmitic (16:0), oleic (18:1 Δ^9) and linoleic (18:2 $\Delta^{9,12}$) acids (Kalac 2013; Pedneault et al. 2006), while several MUFAs and PUFAs including palmitoleic acid (16:1 Δ^9), vaccenic acid (16:1 Δ^{11}), oleic acid (18:1 Δ^9), linoleic acid (18:2 $\Delta^{9,12}$), gondoic acid (20:1 Δ^{11}),

Table 2.5 Essential Fatty Acid Content (% in 100 g of Total Fatty Acids) in Different Species of Mushrooms

Mushroom Species	Linoleic Acid (%)	Linolenic Acid (%)	Oleic Acid (%)	References
Agaricus bisporus	3.3–67.3	1.5–4.6	6.1–55.0	Yilmaz et al. (2006); Özturk et al. (2011); Sadiq et al. (2008); Shoeb et al. (2017)
Boletus edulis	33.6–75.6	0.5–1.7	8.2–31.1	Yilmaz et al. (2006); Abugri et al. (2016); Shao et al. (2010); Yilmaz et al. (2006); Ayaz et al. (2011); Kavishree et al. (2008); Hanus et al. (2008)
Coprinus comatus	25.8–71.1	1.1	5.7–6.2	Yilmaz et al. (2006); Cohen et al. (2014)
Pleurotus ostreatus	4.3–65.29	0.0–6.6	4.8–27.0	Yilmaz et al. (2006); Günç Ergönül et al. (2013); Cohen et al. (2014); Lee et al. (2010); Shoeb et al. (2017); Woldegiorgis et al. (2015); Obodai et al. (2014); Majesty et al. (2019)
Hydnum repandum	27.2–42.5	0.25–20.3	26.4–42.4	Ayaz et al. (2011); Kavishree et al. (2008)
Agaricus bitorquis	31.0–64.4	2.0–11.3	6.9–40.1	Özturk et al. (2011); Sadiq et al. (2008)
Flammulina velutipes	40.9–63.7	1.0–13.0	5.6–16.4	Günç Ergönül et al. (2013); Cohen et al. (2014); Lee et al. (2010); Pereira et al. (2012)
Cantharellus cibarius	17.3–45.6	–	8.4–35.4	Ayaz et al. (2011); Kavishree et al. (2008)

(*Continued*)

Table 2.5 (Continued) Essential Fatty Acid Content (% in 100 g of Total Fatty Acids) in Different Species of Mushrooms

Mushroom Species	Linoleic Acid (%)	Linolenic Acid (%)	Oleic Acid (%)	References
Pleurotus sajor-caju	29.5–53.8	0.2–11.7	16.4–41.7	Kavishree et al. (2008); Atri et al. (2013); Obodai et al. (2014); Kayode et al. (2015)
Termitomyces heimii	35.2–36.0	–	2.9–28.1	Kavishree et al. (2008); Shoeb et al. (2017)
Cordyceps sinensis	25.0	2.4	59.5	Li et al. (2006)
Russula griseocarnosa	28.4	1.0	47.9	Chen et al. (2010)
Tricholoma matsutake	26.7	–	57.6	Liu et al. (2010)
Clitocybe maxima	28.3	0.1	46.8	Liu et al. (2012)
Auricularia polytricha	29.5	3.6	27.1	Kavishree et al. (2008)
Termitomyces microcarpus	2.8–32.7	0.1–8.8	7.4–34.2	Kavishree et al. (2008); Woldegiorgis et al. (2015); Nakalembe & Kabasa (2013)
Lactarius sanguifluus	34.6–35.1	0.70	18.8–32.4	Kalogeropoulos et al. (2013); Kavishree et al. (2008)
Agaricus brasiliensis	65.9	1.10	5.8	Cohen et al. (2014)
Cordyceps militaris	58.7	0.30	11.1	Sande et al. (2019)
Ganoderma lucidum	11.6	1.7	52.4	Sande et al. (2019)
Hericium erinaceus	26.9	0.40	33.7	Sande et al. (2019)
Lentinus edodes	75.8	0.28	3.5	Sande et al. (2019)
Ophiocordyceps sinensis	32.9	2.10	36.3	Sande et al. (2019)

(Continued)

Table 2.5 (Continued) Essential Fatty Acid Content (% in 100 g of Total Fatty Acids) in Different Species of Mushrooms

Mushroom Species	Linoleic Acid (%)	Linolenic Acid (%)	Oleic Acid (%)	References
Trametes versicolor	13.7	1.70	16.2	Sande et al. (2019)
Tremella fuciformis	17.2	0.70	51.8	Sande et al. (2019)
Amanita umbrinolutea	18.8	–	58.8	Pereira et al. (2012)
Endoptychum agaricoides	62.9	0.1	14.5	Marekov et al. (2012)
Hygrocybe nigrescens	18.5	0.0	42.7	Sande et al. (2019)
Laetiporus sulphureus	63.3	0.17	14.5	Petrovic et al. (2014)
Leucoagaricus leucothites	74.7	–	6.3	Pereira et al. (2012)
Russula delica	44.3	2.5	13.7	Kalogeropoulos et al. (2013)
Suillus bellinii	36.7	0.24	39.5	Sande et al. (2019)
Tricholoma portentosum	27.9	–	58.0	Díez and Alvarez (2001)
Tricholoma terreum	29.7	–	56.7	Sande et al. (2019)
Ganoderma applanatum	16.8–17.9	–	32.0–37.4	Abugri et al. (2016); Pedneault et al. (2008)
Macrolepiota procera	47.0	15.6	17.2	Sande et al. (2019)
Lentinus squarrulosus	61.4	1.2	5.8	Sande et al. (2019)
Geastrum arinarius	5.4	1.1	55.1	Sande et al. (2019)
Auricularia polytricha	29.5	3.6	27.1	Kavishree et al. (2008)
Laccaria amethystea	74.4	1.1	13.7	Sande et al. (2019)

(Continued)

Table 2.5 (Continued) Essential Fatty Acid Content (% in 100 g of Total Fatty Acids) in Different Species of Mushrooms

Mushroom Species	Linoleic Acid (%)	Linolenic Acid (%)	Oleic Acid (%)	References
Craterellus cornucopioides	28.4	1.3	54.0	Sande et al. (2019)
Catathelasma ventricosum	73.9	1.0	15.5	Sande et al. (2019)

(*Source:* adapted from Sande et al. 2019)

arachidonic acid (20:4 $\Delta^{5,8,11,14}$), erucic acid (22:1 Δ^{13}) and nervonic acid (24:1 Δ^{15}) are present in both edible wild and cultivated mushroom species (Ribeiro et al. 2009). These edible mushrooms include several species such as *Agaricus arvensis, A. campestris, A. vaginata, B. edulis, Suillus granulates, Amanita rubescens, Boletus subglabripes, Lycoperdon pyriforme, Coprinus comatus, Ganoderma applanatum, Tricholoma* sp., *Fistulina* sp., *Cantharellus* sp. and *Hydnum* sp. (Abugri et al. 2016; Ribeiro et al. 2009; Pedneault et al. 2008; Pedneault et al. 2006). As all wild mushroom species are not edible and may have hallucinogenic and toxic effects, knowledge of edible species is of the utmost importance (Oliveira, Calado and Rosado 2003; Vargas-Isla, Ishikawa and Py-Daniel 2013).

2.4.1.3.3 *Microalgae*

Microalgae are by far the most abundant and initial EPA and DHA producers in the aquatic food chain (Zahringer et al. 2005). With high potential of long-chain omega-3 FA production, these microorganisms can grow naturally fast under several heterotrophic, autotrophic and mixotrophic culture conditions (Li, Qin, Moore and Ball 2009). During photosynthesis, mixotrophic and autotrophic microalgae fix CO_2 from the atmosphere; they also have short growth and harvesting periods and potentially can grow on non-arable lands (Rubio-Rodríguez et al. 2010; Schenk et al. 2008). Microalgal sources are reported having comparatively high EPA and DHA content and productivity in comparison with other potential plant, fish, bacterial and fungal sources (Adarme-Vega et al. 2012).

Massive numbers of marine microalgae species have around 10–50% (w/w) oil content and produce up to 30–70% of total lipid (on dry weight basis) (Ward and Singh 2005). Microalgal cells can produce a number of different compounds for their adaption and survival during unfavorable growth conditions. The production and accumulation of FAs are closely associated with growth stages of microalgae, to function as an energy stockpile during different environmental conditions and cell division (Adarme-Vega et al. 2012). As ω-3 FAs have high energy levels and good flow characteristics, they are found essential for proper functioning of microalgal cells (Cohen et al. 2000; Tiez and Zeiger 2010).

Strains from the genera such as *Thraustochytrium*, *Schizochytrium* and *Ulkenia* from the family Thraustochytriaceae and genus *Cryptocodinium* from the family Crypthecodiniaceae (Borowitzka 2013; Klok et al. 2014), as well as *Phaeodactylum* sp. and *Nannochloropsis* sp., have been demonstrated as rich sources of EPA and/or DHA (Adarme-Vega et al. 2012). When grown heterotrophically, an EPA content of up to 39% of total FAs was accumulated in *Phaeodactylum tricornutum*, *Nannochloropsis* sp., *Nitzchia* sp. and *Monodus* sp. (Wen and Chen 2003; Ji et al. 2015; Chua and Schenk 2017; Steinrucken et al. 2017; Chen et al. 2018), while a DHA percentage around 30–40% of total FAs was reported for strains such as *Thraustochytrium* sp. (Burja et al. 2006), *Schizochytrium limacinum* (Zhu et al. 2007), *S. mangrovei* (Chen et al. 2007) and *Cryptocodinium cohnii* (Kyle 2001). Currently, the use of *Schizochytrium* sp. has been reported in various food processing industries and food products (Sijtsma and de Swaaf 2004), while *C. cohnii* is being used in different commercial products (Kyle 2001). High microalgal biomass, and high oil and EPA and DHA productivities are achieved with media having optimized carbon and nitrogen levels, controlled environmental conditions such as oxygen, pH, temperature and controlled nutrient supply including glucose, sodium, nitrogen, etc. (Griffiths and Harrison 2009; Ward and Singh 2005).

Although algal oils are produced in a carefully controlled environment, are suitable for vegetarian diets and have excellent sustainability properties (Winwood 2013), some limiting factors such as cost and extraction and purification methods are currently restricting their potential for use on a larger scale (Adarme-Vega et al. 2012).

2.4.1.4 Issues with Current Sources: Need for Alternative Sources of Omega-3 Fatty Acids (EPA and DHA)

Due to the apparent limitations of larger organisms, including humans, for synthesizing omega-3 fatty acids, seafood is the principal nutritional source for humans (Swanson, Block and Mousa 2012). The Food and Agriculture Organization of the United Nations in its published data suggested that the supply of fish is steadily increasing worldwide by 3.2% per year (FAO 2014). However, high demand for fish oil, decline in global fish stocks and compromises to biodiversity by destruction of wild fish habitats and natural marine processes, overfishing, pollution and metal toxin accumulation in fish stock (for example, by mercury), as well as demand for vegan long-chain PUFA (LC-PUFA) sources, are some of the major factors calling for an alternative source for these EFAs (Adarme-Vega et al. 2012). High cost and investment, as well as reduced production of microalgal ω-3 FAs in response to global warming, have added an impending decline in overall ω-3 PUFA production (Gladyshev, Arts and Sushchik 2009). Moreover, the biosynthetic pathway in plants is unable to synthesize LC ω-3 FAs with many double bonds (such as EPA and DHA) of greater nutritional interest (Haslam et al. 2013). These current challenges and concerns have created the human dietary supply of ω-3 PUFAs as a relevant issue to be addressed. Therefore, it is the need of the hour to find some cost-effective ways to produce abundant as well as safe sources for fulfilling the demands of PUFA-rich oils.

2.4.1.5 Genetically Engineered Plants and Microorganisms

To overcome the above-mentioned challenges and limitations, alternative approaches such as the modification of the genetic makeup of higher plants and microorganisms have been attempted (Table 2.6). Scientists have emphasized the effectiveness of ω-3 FA production through genetic manipulation of plants (Ruiz-Lopez et al. 2012; Chen, Meesapyodsuk and Qiu 2014). In higher plants, multiple genes of the PUFA biosynthesis pathway derived from other sources, such as fungi, bacteria, microalgae and diatoms, were introduced. As a result, a number of different transgenic plants (e.g., *Brassica*, soybean, *Arabidopsis*, tobacco, etc.) with the capability of producing attractive yields of DHA and EPA (Haslam et al. 2013; Napier and Graham 2010; Rogalski and Carrer 2011; Sayanova and Napier 2011) have been accomplished. To produce fair quantities of ω-3 FAs, the scientists focused on manipulation in the biosynthesis pathway of higher plants. Three microbial enzymes, namely *ΔD9 elongase* (*Isochrysis galbana*), *ΔD8 desaturase* (*Euglena gracilis*) and a *D5 desaturase* (*Mortierella alpina*), were overexpressed in the leaf tissues of *Arabidopsis thaliana*, which enhanced the biosynthesis of ARA and EPA by 7% and 3%, respectively (Barclay et al. 1994). Also, Wu et al. (2005a, 2005b) later reconstituted the entire DHA biosynthesis pathway in an oilseed crop, *Brassica juncea*, through stepwise metabolic engineering and produced considerable AA (25%), EPA (15%) and DHA (1.5%) in the seeds. Similarly, the genetic engineering of *Camelina sativa* was established by inserting 5–7 genes in the FA pathway,

Table 2.6 Genetic Engineering Studies of Transgenic Plants and Microorganisms to Improve PUFA Yield

Source	Type	Target Fatty Acid	References
Plants	Plant	EPA and DHA	Wu et al. (2005a, b)
Mustard	Plant	EPA and DHA	Kinney et al. (2004)
Brassica carinata	Plant	EPA and DHA	Petrie et al. (2010)
Nicotiana benthamiana	Plant	EPA and DHA	Petrie et al. (2010)
Camelina sativa	Plant	EPA and DHA	Ruiz-Lopez, Haslam, Napier and Sayanova (2014); Petrie, Shrestha, Belide, Kennedy and Lester (2014)
Aspergillus oryzae	Fungi	PUFA	Sakuradani et al. (2013)
Pichia pastoris	Fungi	EPA	Li et al. (2009); Jiang et al. (2014); Kajikawa et al. (2004)
Yarrowia lipolytica	Fungi	EPA, PUFA	Chuang et al. (2010); Xue et al. 2013);
Mortierella alpina CFR-GV15	Fungi	EPA	Vadivelan and Venkateswaran (2014)
Lobosphaera incisa	Microalgae	DGLA (Dihomo-g-linolenic acid)	Abu-Ghosh et al. (2015)
Phaeodactylum tricornutum	Microalgae	DHA	Hamilton et al. (2014)
Mortierella alpine 1S-4	Fungi	AA, DGLA (Dihomo-g-linolenic acid)	Sakuradani et al. (2013)

which resulted in accumulation of 12% EPA and 14% DHA, which was similar to fish (Ruiz-Lopez et al. 2014). Analogously, many obstacles such as control of the synthesis intermediates released by different organelles and reservoirs, control of the primary synthesis, control of flux of substrates and biosynthetic precursors, reduction in the stimulation of the ω-6 branch of the pathway, etc., have also been accomplished (Zárate, el Jaber-Vazdekis and Ramírez-Moreno 2016). Native oleic acid, LA and ALA were also converted mainly into EPA and DHA in the engineered species of *Camelina sativa* (Petrie et al. 2014). More importantly, a total of 25% ω-3 FAs (eicosatetraenoic acid (ETA, 20:4n-3), EPA and DPA) with up to 12% DHA and only 1.2–1.4% of GLA as well as no other omega-6 PUFAs were recorded. In tobacco, significant AA (15.5%) and EPA (4.9%) were accumulated in tobacco after insertion of some genes from *Marchantia polymorpha* (Kajikawa et al. 2008). Although the results demonstrated the effectiveness of these genetically modified (GM) higher plants in offering a sustainable and real agronomic alternative source for LC ω-3 PUFA production, but before implementing these GM crops for FA production, several issues regarding the acceptance of and opposition to these crops have to be overcome (Zárate, el Jaber-Vazdekis and Ramírez-Moreno 2016).

Different microbial host systems including microalgae, fungi and bacteria have been investigated to gain insight into the ω-3 biosynthesis pathway. For example, yeast *Yarrowia lipolytica* was genetically engineered by E. I. DuPont Company, which is used for commercial mass-scale production of EPA (Xie et al. 2015). Transgenic *Escherichia coli*, producing 6.1 times higher EPA compared to marine bacterium *Shewanella baltica* MAC1, was produced by successful cloning of a gene cluster (capable to produce EPA and DHA) from *S. baltica* MAC1 strain found in fish intestine (Amiri-Jami and Griffiths 2010). Also, in another investigation, gene clusters from *S. baltica* MAC1 were expressed in *Lactococcus lactis* subsp. *cremoris* MG1363, which resulted in high yields of omega-3 PUFA (EPA and DHA) (Amiri-Jami et al. 2014).

2.4.1.6 Fortified Foods

In addition to the natural dietary sources of EPA and DHA, the market for industrialized foods enriched with omega-3 PUFA, especially DHA, is growing. Foods can be enriched with n-3 PUFAs by 1) enriching animal tissues and dairy products by modification of the animal diet using n-3 PUFA-rich products; 2) microencapsulated/ modified oil obtained from transgenic microbial and plant sources; and 3) fortification of different foods such as baked goods, infant formula, etc., with plant or microalgal oils (Arbex et al. 2015). Nowadays, a variety of omega-3 PUFA-fortified food products such as bread and bakery products, poultry, milk and dairy products, egg products, soft drinks, juices and different spreads have been used to improve the human dietary status of n-3 PUFAs (Ganesan et al. 2014; Kolanowski and Laufenberg 2006; Meluzzi et al. 2000; Shapira, Weill and Loewenbach 2008). Some heterotrophic microalgae are in particular added to infant milk formula and other foods, while other microalgal products may be used as animal feed, food additives, pharma products, etc. (Batista et al. 2007; Puri 2017; Koyande et al. 2019).

2.4.2 Omega-6 Fatty Acids

Omega-6 FAs can be obtained from different whole grains, wheat germ, dried fruits and nuts including walnuts, almonds, pistachios, Brazil nuts and peanuts (Meyer et al. 2003; Martins et al. 2008). A few seeds, including sesame seeds, chia seeds, safflower seeds, hemp seeds, grape seeds, pumpkin seeds and their oils can also fulfill the dietary requirement for linoleic acids (Meyer et al. 2003; Martins et al. 2008; Bimbo 2013; Johnson and Bradford 2014). Meat, poultry, eggs and all types of fish (e.g., mackerel, salmon and tuna) are some rich sources of omega-6 FAs, especially arachidonic acid, for the non-vegetarian platter (Meyer et al. 2003) (Table 2.7).

Although gamma-linolenic acid (GLA), a conditionally essential FA, can be produced in the body from linoleic acid, in some circumstances, the body cannot convert LA to GLA efficiently. These include diabetes, old age, high consumption of alcohol and/or saturated fat, high cholesterol level, viral infection and deficiencies of some essential minerals and vitamins (Paola Benatti et al. 2004). In such conditions, dietary GLA supplementation with some sources such as borage oil (richest source with 17% to 25%), black currant oil (15% to 20%) and evening primrose oil (7% to 10%) is used.

A balance between the supplementation of omega-3 (EPA and DHA) and omega-6 (GLA) is necessary for people who wish to increase their dietary fatty acid. Equazen provides a balanced supplementation of these essential PUFAs with a scientifically proven ratio of EPA:DHA:GLA equal to 9:3:1, which resulted in improved attention and learning, as well as reduced hyperactivity, in clinical trials (Gómez Candela et al. 2011).

2.4.3 Omega-9 Fatty Acids

An omega-9 FA, oleic acid (C18:1n-9), has been suggested to contribute to the metabolism of the EFAs (Dhopeshwarkar and Mead 1961; Lowry and Tinsley 1966) as bioactive lipid mediators. By stimulating the enzymes and the production of cytokines and other molecules, these FAs regulate pro- and anti-inflammatory processes (Maskrey et al. 2011). Olive oil and macadamia oil are the rich dietary sources of oleic acid, whereas nervonic acid (C24:1n-9), another n-9 FA, is mainly found in king salmon, yellow

Table 2.7 Food Sources for Omega-6 Fatty Acids

Food Groups	**Linoleic Acid, 18:2 (%)**	**Eicosadienoic Acid, 20:2 (%)**	**Eicosatrienoic Acid, 20:3 (%)**	**Arachidonic Acid, 20:4 (%)**	**Adrenic Acid, 22:4 (%)**	**∑n-6 (Sum of 20:5, 22:5 and 22:6) (%)**
Meat and poultry products	15.26	20.58	49.21	11.77	19.6	15.86
Cereal-based products including biscuits, cakes, pies, pizza, pancakes, etc.	23.8	0.15	0.40	4.35	2.55	20.85
Confectionery products and health bars	0.53	0	0	0	0	0.46
Milk and other dairy products	11.40	0	0	0	0	9.82
Vegetables and their products	9.81	2.22	0	0	0.66	8.54
Fats and oils	16.31	0	0	0	0	14.05
Cereals and cereal products including bread, muffins, rice, pasta and breakfast cereals	12.86	0.05	0	0.33	0.20	11.10
Fish and seafood products	3.54	75.87	50.36	73.12	70.78	12.86
Egg and egg products	0.55	1.12	0.04	10.02	5.98	1.30
Savory sauces and condiments	3.54	0	0	0.37	0.21	3.08
Legumes and pulses	1.01	0	0	0.03	0.02	0.87
Seeds and nuts	0.43	0	0	0	0	0.37

(*Source:* adapted from Meyer et al. 2003)

mustard seed and flaxseed (Johnson and Bradford 2014). These FAs independently yet synergistically act with other bio-molecules within the cellular ecosystem and help in maintaining homeostasis and overall health.

2.5 CONCLUSIONS

Much evidence is available supporting the benefits of EFAs to human health, and the focus has been on preventing several chronic diseases, especially reducing cardiovascular, cognitive and eye health problems. Recent studies have suggested special benefits to the pregnancy and childhood stages, which further emphasized consistent and balanced dietary supply of these EFAs. For better conversion of ALA to DHA and maximum possible health benefits, current recommendations suggest an n-6 to n-3 ratio of 2:1 to 3:1 in human diets, but due to industrialization, this is highly imbalanced in modern diets.

The parent ω-3 FA (i.e., ALA) can be obtained primarily from seeds and their oils such as flaxseeds, perilla, chia and echium oil. Among animal sources, marine organisms, especially fish such as mackerel, herring and salmon, some oily fishes such as *Scombridae*, *Clupeidae* and *Salmonidae* families, other seafood, red and white meats, as well as egg yolk are main sources of different ω-3 PUFAs. Compared to marine fishes, high FA content was found in freshwater fish which could be due to the high FA composition of freshwater phytoplankton. Oleaginous marine bacteria, algae and fungi are some microbial sources of ω-3 PUFAs. Short life cycle, the availability of cheap and abundant raw material and easy large-scale production make microbial oil production advantageous over other sources. Mushrooms, although low in lipid content, contain PUFAs (LA, oleic acid and ALA) as major constituents of their lipid profile. As wild species may have hallucinogenic and toxic effects, knowledge of edible mushroom species is essential. Microalgal sources are the most abundant and initial EPA and DHA producers in aquatic food chains, having high potential for LC ω-3 FA production in comparison with other available plant, fish, bacterial and fungal sources. The algal oils are suitable for vegetarian diets and have excellent sustainability properties but production cost and extraction and purification methods are some limiting factors for their large-scale use.

Factors such as high fish oil demand, decreasing global fish stocks, overfishing, pollution and environmental pollutants accumulated in fish as well as demand for vegan EFAs sources limit the use of current sources and demand for new novel alternative sources of these EFAs. Alternative approaches such as transgenic derivation and extensive metabolic engineering of plants and microorganisms can be established to produce sustainable and cost-effective alternative sources for fish oils. To overcome the above-mentioned challenges and limitations, alternative approaches such as modification of the genetic makeup of higher plants and microorganisms may be used. Fortified foods can also be one alternative to correct the omega fatty acid imbalance, but considering the interaction of ω-3 PUFAs with other food ingredients, its impact on the nutritional effects of ω-3 PUFAs deserves further investigation.

REFERENCES

Abu-Ghosh, S., Pal-Nath, D., Markovitch, D., Solovchenko, A., Didi-Cohen, S., Portugal, I., Khozin-Goldberg, I., Cohen, Z., and Boussiba, S. 2015. A novel source of dihomo-γ-linolenic acid: Possibilities and limitations of DGLA production in the high-density cultures of the Δ5 desaturase-mutant microalga *Lobosphaera incisa*. *European Journal of Lipid Science and Technology* 117(6): 760–766.

Abugri, D., McElhenney, W. H., and Willian, K. R. 2016. Fatty acid profiling in selected cultivated edible and wild medicinal mushrooms in the Southern United States. *Journal of Experimental and Food Chemistry* 2: 1–7.

Adarme-Vega, T. C., Lim, D. K. Y., Timmins, M., Vernen, F., Li, Y., and Schenk, P. M. 2012. Microalgal biofactories: A promising approach towards sustainable omega-3 fatty acid production. *Microbial Cell Factories* 11(1): 86–96.

Allemann, M. N., Shulse, C., and Allen, E. E. 2019. Linkage of marine bacterial polyunsaturated fatty acid and long-chain hydrocarbon biosynthesis. *Frontiers in Microbiology* 10: 702.

Ambring, A., Johansson, M., Axelsen, M., Gan, L., Strandvik, B., and Friberg, P. 2006. Mediterranean-inspired diet lowers the ratio of serum phospholipid n-6 to n-3 fatty acids, the number of leukocytes and platelets, and vascular endothelial growth factor in healthy subjects. *The American Journal of Clinical Nutrition* 83(3): 575–581.

Amiri-Jami, M., and Griffiths, M. W. 2010. Recombinant production of omega-3 fatty acids in Escherichia coli using a gene cluster isolated from *Shewanella baltica* MAC1. *Journal of Applied Microbiology* 109(6): 1897–1905.

Amiri-Jami, M., LaPointe, G., and Griffiths, M. W. 2014. Engineering of EPA/DHA omega-3 fatty acid production by *Lactococcus lactis subsp. cremoris* MG1363. *Applied Microbiology and Biotechnology* 98(7): 3071–3080.

Amjad Khan, W., Chun-Mei, H., Khan, N., Iqbal, A., Lyu, S. W., and Shah, F. 2017. Bioengineered plants can be a useful source of omega-3 fatty acids. *BioMed Research International* 9: 7348919.

Ando, Y., Nishimura, K., Aoyanagi, N., and Takagi, T. 1992. Stereospecific analysis of fish oil triacyl-sn-glycerols. *Journal of the American Oil Chemists' Society* 69(5): 417–424.

Arbex, A. K., Bizarro, V. R., Santos, J. C. S., Araújo, L. M. M., de Jesus, A. L. C., Fernandes, M. S. A., Salles, M. M., Rocha, D. R. T. W., and Marcadenti, A. 2015. The impact of the essential fatty acids (EFA) in human health. *Open Journal of Endocrine and Metabolic Diseases* 5(07): 98–104.

Arterburn, L. M., Hall, E. B., and Oken, H. 2006. Distribution, interconversion, and dose response of n-3 fatty acids in humans. *The American Journal of Clinical Nutrition* 83(6): 1467S–1476S.

Athalye, S. K., Garcia, R. A., and Wen, Z. 2009. Use of biodiesel-derived crude glycerol for producing eicosapentaenoic acid (EPA) by the fungus *Pythium irregulare. Journal of Agricultural and Food Chemistry* 57(7): 2739–2744.

Atri, N. S., Sharma, S. K., Joshi, R., Gulati, A., and Gulati, A. 2013. Nutritional and neutraceutical composition of five wild culinary-medicinal species of genus *Pleurotus* (higher Basidiomycetes) from northwest India. *International Journal of Medicinal Mushrooms* 15(1): 49–56.

Ayaz, F. K. A., Chuang, L. T., Torun, H., Colak, A., Sesli˙, E., Presley, J., Smith, B. R., and Glew, R. H. 2011. Fatty acid and amino acid compositions of selected wild-edible mushrooms consumed in Turkey. *International Journal of Food Sciences and Nutrition* 62(4): 328–335.

Bailey, N. 2009. Current choices in omega 3 supplementation. *Nutrition Bulletin* 34(1): 85–91.

Barclay, W. R., Meager, K. M., and Abril, J. R. 1994. Heterotrophic production of long chain omega-3 fatty acids utilizing algae and algae-like microorganisms. *Journal of Applied Phycology* 6(2): 123–129.

Batista, A. P., Bandarra, N., Raymundo, A., and Gouveia, L. 2007. Microalgae biomass-a potential ingredient for the food industry. In: *EFFoST/EHED Joint Conference,* Lisbon, Portugal.

Benatti, P., Peluso, G., Nicolai, R., and Calvani, M. 2004. Polyunsaturated fatty acids: Biochemical, nutritional and epigenetic properties. *Journal of the American College of Nutrition* 23(4): 281–302.

Bergé, J. P., and Barnathan, G. 2005. Fatty acids from lipids of marine organisms: Molecular biodiversity, roles as biomarkers, biologically active compounds, and economical aspects. In: Ulber, R. and Le Gal, Y. (eds), *Marine Biotechnology I. Advances in Biochemical Engineering/Biotechnology,* vol. 96, 49–125. Springer: Berlin, Heidelberg.

Bhosale, R. A., Rajabhoj, M. P., and Chaugule, B. B. 2010. *Dunaliella salina Teod.* as a prominent source of eicosapentaenoic acid. *International Journal on Algae* 12(2): 185–189.

Bimbo, A. P. 2013. Sources of omega-3 fatty acids. In: Jacobsen, C., Nielsen, N. S., Horn, A. F. and Sørensen, A.M. (eds), *Food Enrichment with Omega-3 Fatty Acids,* 27–107. Woodhead Publishing, Cambridge, UK.

Borowitzka, M. A. 2013. High-value products from microalgae-their development and commercialisation. *Journal of Applied Phycology* 25(3): 743–756.

Bourdon, J. A., Bazinet, T. M., Arnason, T. T., Kimpe, L. E., Blais, J. M., and White, P. A. 2010. Polychlorinated biphenyls (PCBs) contamination and aryl hydrocarbon receptor (AhR) agonist activity of Omega-3 polyunsaturated fatty acid supplements: Implications for daily intake of dioxins and PCbs. *Food and Chemical Toxicology: An International Journal Published for the British Industrial Biological Research Association* 48(11): 3093–3097.

Bourre, J. M. 2007. Dietary omega-3 fatty acids for women. *Biomedicine and Pharmacotherapy = Biomedecine and Pharmacotherapie* 61(2–3): 105–112.

Breivik, H. (Ed.) 2007. *Long-Chain Omega-3 Specialty Oils* (Vol. 21). Oily Press Lipid Library.

Burdge, G. C., and Calder, P. C. 2005. Conversion of $\alpha $-linolenic acid to longer-chain polyunsaturated fatty acids in human adults. *Reproduction, Nutrition, Development* 45(5): 581–597.

Burja, A. M., Radianingtyas, H., Windust, A., and Barrow, C. J. 2006. Isolation and characterization of polyunsaturated fatty acid producing *Thraustochytrium* species: Screening of strains and optimization of omega-3 production. *Applied Microbiology and Biotechnology* 72(6): 1161–1169.

Candela, C. G., López, L. B., and Kohen, V. L. 2011. Importance of a balanced omega 6/omega 3 ratio for the maintenance of health. Nutritional recommendations. *Nutricion hospitalaria* 26(2): 323–329.

Carvalho, A. P., Meireles, L. A., and Malcata, F. X. 2006. Microalgal reactors: A review of enclosed system designs and performances. *Biotechnology Progress* 22(6): 1490–1506.

Chaturvedi, R., and Fujita, Y. 2006. Isolation of enhanced eicosapentaenoic acid producing mutants of *Nannochloropsis oculata* ST-6 using ethyl methane sulfonate induced mutagenesis techniques and their characterization at mRNA transcript level. *Phycological Research* 54(3): 208–219.

Chen, C. Y., Nagarajan, D., and Cheah, W. Y. 2018. Eicosapentaenoic acid production from *Nannochloropsis oceanica* CY2 using deep sea water in outdoor plastic-bag type photobioreactors. *Bioresource Technology* 25(3): 1–7.

Chen, Y., Meesapyodsuk, D., and Qiu, X. 2014. Transgenic production of omega-3 very long chain polyunsaturated fatty acids in plants: Accomplishment and challenge. *Biocatalysis and Agricultural Biotechnology* 3: 38–43.

Chen, Y. Q., Edwards, I. J., Kridel, S. J., Thornburg, T., and Berquin, I. M. 2007. Dietary fat–gene interactions in cancer. *Cancer and Metastasis Reviews* 26(3–4): 535–551.

Cheng, B., Wu, G., Vrinten, P., Falk, K., Bauer, J., and Qiu, X. 2010. Towards the production of high levels of eicosapentaenoic acid in transgenic plants: The effects of different host species, genes and promoters. *Transgenic Research* 19(2): 221–229.

Chua, E. T., and Schenk, P. M. 2017. A biorefinery for Nannochloropsis: Induction, harvesting, and extraction of EPA-rich oil and high-value protein. *Bioresource Technology* 244: 1416–1424.

Chuang, L. T., Chen, D. C., Nicaud, J. M., Madzak, C., Chen, Y. H., and Huang, Y. S. 2010. Co-expression of heterologous desaturase genes in *Yarrowia lipolytica*. *New Biotechnology* 27(4): 277–282.

Cohen, N., Cohen, J., Asatiani, M. D., Varshney, V. K., Yu, H. T., Yang, Y. C., ... Wasser, S. P. 2014. Chemical composition and nutritional and medicinal value of fruit bodies and submerged cultured mycelia of culinary-medicinal higher basidiomycetes mushrooms. *International Journal of Medicinal Mushrooms* 16(3): 273–291.

Cohen, Z., Khozin-Goldberg, I., Adlerstein, D., and Bigogno, C. 2000. The role of triacylglycerol as a reservoir of polyunsaturated fatty acids for the rapid production of chloroplastic lipids in certain microalgae. *Biochemical Society Transactions* 28(6): 740–743.

Cordova, L. T., and Alper, H. S. 2018. Production of α-linolenic acid in *Yarrowia lipolytica* using low-temperature fermentation. *Applied Microbiology and Biotechnology* 102(20): 8809–8816.

de Oliveira, M. R., Nabavi, S. F., Daglia, M., Rastrelli, L., and Nabavi, S. M. 2016. Epigallocatechin gallate and mitochondria-a story of life and death. *Pharmacological Research* 104: 70–85.

Dhopeshwarkar, G. A., and Mead, J. F. 1961. Role of oleic acid in the metabolism of essential fatty acids. *Journal of the American Oil Chemists' Society* 38(6): 297–301.

Dıez, V. A., and Alvarez, A. 2001. Compositional and nutritional studies on two wild edible mushrooms from northwest Spain. *Food Chemistry* 75(4): 417–422.

Domenichiello, A. F., Kitson, A. P., and Bazinet, R. P. 2015. Is docosahexaenoic acid synthesis from α-linolenic acid sufficient to supply the adult brain? *Progress in Lipid Research* 59: 54–66.

Doughman, S. D., Krupanidhi, S., and Sanjeevi, C. B. 2007. Omega-3 fatty acids for nutrition and medicine: Considering microalgae oil as a vegetarian source of EPA and DHA. *Current Diabetes Reviews* 3(3): 198–203.

Ergönül, P. G., Akata, I., Kalyoncu, F., and Ergönül, B. 2013. Fatty acid compositions of six wild edible mushroom species. *The Scientific World Journal*, 1–4.

Ethier, S., Woisard, K., Vaughan, D., and Wen, Z. 2011. Continuous culture of the microalgae *Schizochytrium limacinum* on biodiesel-derived crude glycerol for producing docosahexaenoic acid. *Bioresource Technology* 102(1): 88–93.

Fan, K. W., Jiang, Y., Faan, Y. W., and Chen, F. 2007. Lipid characterization of mangrove thraustochytrid–*Schizochytrium mangrovei*. *Journal of Agricultural and Food Chemistry* 55(8): 2906–2910.

FAO. 1994. Fats and oils in human nutrition. Report of a Joint Expert Consultation, 168. Food and Agriculture Organization of the United Nations: Rome, Italy.

FAO. 2014. The state of world fisheries and aquaculture. Opportunities and challenges. Food and Agriculture Organization of the United Nations: Rome, Italy.

Fernández, P. M., and Juan, S. 2000. Fatty acid composition of commercial Spanish fast food and snack food. *Journal of Food Composition and Analysis* 13(3): 275–281.

Furlan, V. J. M., Paulo, M. D. C., Maus, V. W., Ferreira, J., Batista, I., and Bandarra, N. M. 2017. Production of docosahexaenoic acid (DHA) from Thraustochytrium sp. ATCC 26185 using differents nitrogen concentrations. *Boletim do centro de pesquisa de processamento de alimentos* 32(1): 1–10.

Ganesan, B., Brothersen, C., and McMahon, D. J. 2014. Fortification of foods with omega-3 polyunsaturated fatty acids. *Critical Reviews in Food Science and Nutrition* 54(1): 98–114.

Gayathri, S., Prabhu, S., Mitra, A., Kumar, V., Sheela, J. S., Rajani, G., ... Sivvaswamy, S. N. 2010. Soil microorganisms produce omega-3 fatty acids. *Indian Journal of Science and Technology* 3(5): 499–503.

Gladyshev, M. I., Arts, M. T., and Sushchik, N. I. 2009. Preliminary estimates of the export of omega-3 highly unsaturated fatty acids (EPA+DHA) from aquatic to terrestrial ecosystems. In: Kainz, M., Brett, M., and Arts, M. (eds), *Lipids in Aquatic Ecosystems*, 179–210. Springer: New York, NY.

Griffiths, M. J., and Harrison, S. T. 2009. Lipid productivity as a key characteristic for choosing algal species for biodiesel production. *Journal of Applied Phycology* 21(5): 493–507.

Guihéneuf, F., Mimouni, V., Ulmann, L., and Tremblin, G. 2009. Combined effects of irradiance level and carbon source on fatty acid and lipid class composition in the microalga *Pavlova lutheri* commonly used in mariculture. *Journal of Experimental Marine Biology and Ecology* 369(2): 136–143.

Günç Ergönül, P., Akata, I., Kalyoncu, F., and Ergönül, B. 2013. Fatty acid compositions of six wild edible mushroom species. *The Scientific World Journal* 2013: 1–4. doi:10.1155/2013/163964.

Gupta, A., Barrow, C. J., and Puri, M. 2012. Omega-3 biotechnology: Thraustochytrids as a novel source of omega-3 oils. *Biotechnology Advances* 30(6): 1733–1745.

Hamilton, M. L., Haslam, R. P., Napier, J. A., and Sayanova, O. 2014. Metabolic engineering of *Phaeodactylum tricornutum* for the enhanced accumulation of omega-3 long chain polyunsaturated fatty acids. *Metabolic Engineering* 22: 3–9.

Hamilton, M. L., Powers, S., Napier, J. A., and Sayanova, O. 2016. Heterotrophic production of omega-3 long-chain polyunsaturated fatty acids by trophically converted marine diatom *Phaeodactylum tricornutum*. *Marine Drugs* 14(3): 53.

Hanus, L. O., Shkrob, I., and Dembitsky, V. M. 2008. Lipids and fatty acids of wild edible mushrooms of the genus *Boletus*. *Journal of Food Lipids* 15(3): 370–383.

Hanzel, J. R., and Williams, E. E. 1990. The role of alteration in membrane lipid composition in enabling physiological adaptation of organisms to their physical environment. *Progress in Lipid Research* 29(3): 167–227.

Haslam, R. P., Ruiz-Lopez, N., Eastmond, P., Moloney, M., Sayanova, O., and Napier, J. A. 2013. The modification of plant oil composition via metabolic engineering-better nutrition by design. *Plant Biotechnology Journal* 11(2): 157–168.

Hu, C., Li, M., Li, J., Zhu, Q., and Liu, Z. 2008. Variation of lipid and fatty acid compositions of the marine microalga *Pavlova viridis* (*Prymnesiophyceae*) under laboratory and outdoor culture conditions. *World Journal of Microbiology and Biotechnology* 24(7): 1209–1214.

Hu, H., and Gao, K. 2003. Optimization of growth and fatty acid composition of a unicellular marine picoplankton, *Nannochloropsis* sp., with enriched carbon sources. *Biotechnology Letters* 25(5): 421–425.

Islam, M. S., Castellucci, C., Fiorini, R., Greco, S., Gagliardi, R., Zannotti, A., Giannubilo, S. R., Ciavattini, A., Frega, N. G., Pacetti, D., and Ciarmela, P. 2018. Omega-3 fatty acids modulate the lipid profile, membrane architecture, and gene expression of leiomyoma cells. *Journal of Cellular Physiology* 233(9): 7143–7156.

Jareonkitmongkol, S., Shimizu, S., and Yamada, H. 1993. Production of an eicosapentaenoic acid-containing oil by a Δ12 desaturase-defective mutant of *Mortierella alpina* 1S-4. *Journal of the American Oil Chemists' Society* 70(2): 119–123.

Jermsuntiea, W., Aki, T., Kawamoto, S., and Ono, K. 2011. Metabolism and synthesis of lipids in the polyunsaturated fatty acid-producing fungus *Mortierella alliacea*. *Journal of Oleo Science* 60(1): 11–17.

Ji, X. J., Ren, L. J., and Huang, H. 2015. Omega-3 biotechnology: A green and sustainable process for omega-3 fatty acids production. *Frontiers in Bioengineering and Biotechnology* 3: 158.

Jiang, M., Guo, B., Wan, X., Gong, Y., Zhang, Y., and Hu, C. 2014. Isolation and characterization of the diatom *Phaeodactylum* Δ5-elongase gene for transgenic LC-PUFA production in *Pichia pastoris*. *Marine Drugs* 12(3): 1317–1334.

Johnson, M., and Bradford, C. 2014. Omega-3, omega-6 and omega-9 fatty acids: Implications for cardiovascular and other diseases. *Journal of Glycomics and Lipidomics* 4(123): 2153–0637.

Kajikawa, M., Matsui, K., Ochiai, M., Tanaka, Y., Kita, Y., Ishimoto, M., Kohzu, Y., Shoji, S., Yamato, K. T., Ohyama, K., Fukuzawa, H., and Fukuzawa, H. 2008. Production of arachidonic and eicosapentaenoic acids in plants using bryophyte fatty acid Δ6-desaturase, Δ6-elongase, and Δ5-desaturase genes. *Bioscience, Biotechnology, and Biochemistry* 72(2): 435–444.

Kajikawa, M., Yamato, K. T., Kohzu, Y., Nojiri, M., Sakuradani, E., Shimizu, S., Sakai, Y., Fukuzawa, H., and Ohyama, K. 2004. Isolation and characterization of Δ 6-desaturase, an ELO-like enzyme and Δ 5-desaturase from the liverwort *Marchantia polymorpha* and production of arachidonic and eicosapentaenoic acids in the methylotrophic yeast *Pichia pastoris*. *Plant Molecular Biology* 54(3): 335–352.

Kalač, P. 2013. A review of chemical composition and nutritional value of wild-growing and cultivated mushrooms. *Journal of the Science of Food and Agriculture* 93(2): 209–218.

Kalogeropoulos, N., Yanni, A. E., Koutrotsios, G., and Aloupi, M. 2013. Bioactive microconstituents and antioxidant properties of wild edible mushrooms from the island of Lesvos, Greece. *Food and Chemical Toxicology: An International Journal Published for the British Industrial Biological Research Association* 55: 378–385.

Kamei, M., Ki, M., Kawagoshi, M., and Kawai, N. 2002. Nutritional evaluation of Japanese take-out lunches compared with Western-style fast foods supplied in Japan. *Journal of Food Composition and Analysis* 15(1): 35–45.
Kato, C., and Nogi, Y. 2001. Correlation between phylogenetic structure and function: Examples from deep-sea *Shewanella*. *FEMS Microbiology Ecology* 35(3): 223–230.
Kaur, N., Chugh, V., and Gupta, A. K. 2014. Essential fatty acids as functional components of foods: A review. *Journal of Food Science and Technology* 51(10): 2289–2303.
Kavishree, S., Hemavathy, J., Lokesh, B. R., Shashirekha, M. N., and Rajarathnam, S. 2008. Fat and fatty acids of Indian edible mushrooms. *Food Chemistry* 106(2): 597–602.
Kayode, R. M. O., Olakulehin, T. F., Adedeji, B. S., Ahmed, O., Aliyu, T. H., and Badmos, A. H. A. 2015. Evaluation of amino acid and fatty acid profiles of commercially cultivated oyster mushroom (*Pleurotus sajor-caju*) grown on gmelina wood waste. *Nigerian Food Journal* 33(1): 18–21.
Khozin-Goldberg, I., Didi-Cohen, S., Shayakhmetova, I., and Cohen, Z. 2002. Biosynthesis of eicosapentaenoic acid (EPA) in the freshwater eustigmatophyte *Monodus subterraneus* (*Eustigmatophyceae*) 1. *Journal of Phycology* 38(4): 745–756.
Kinney, A. J., Cahoon, E. B., Damude, H. G., Hitz, W. D., Kolar, C. W., and Liu, Z. 2004. Production of very long chain polyunsaturated fatty acids in oilseed plants. International Patent Application WO 2004/071467 A2.
Klok, A. J., Lamers, P. P., Martens, D. E., Draaisma, R. B., and Wijffels, R. H. 2014. Edible oils from microalgae: Insights in TAG accumulation. *Trends in Biotechnology* 32(10): 521–528.
Kolanowski, W., and Laufenberg, G. 2006. Enrichment of food products with polyunsaturated fatty acids by fish oil addition. *European Food Research and Technology* 222(3–4): 472–477.
Koyande, A. K., Chew, K. W., Rambabu, K., Tao, Y., Chu, D. T., and Show, P. L. 2019. Microalgae: A potential alternative to health supplementation for humans. *Food Science and Human Wellness* 8(1): 16–24. doi:10.1016/j.fshw.2019.03.001.
Kris-Etherton, P. M., Taylor, D. S., Yu-Poth, S., Huth, P., Moriarty, K., Fishell, V., Hargrove, R. L., Zhao, G., and Etherton, T. D. 2000. Polyunsaturated fatty acids in the food chain in the United States. *American Journal of Clinical Nutrition* 71: 179S–188S.
Kyle, D. J. 2001. The large-scale production and use of single-cell oil highly enriched in docosahexaenoic acid. *American Chemical Society* 8: 92–107.
Lee, J. Y., Yoo, C., Jun, S. Y., Ahn, C. Y., and Oh, H. M. 2010. Comparison of several methods for effective lipid extraction from microalgae. *Bioresource Technology* 101(1): S75–S77.
Lenihan-Geels, G., Bishop, K. S., and Ferguson, L. R. 2013. Alternative sources of omega-3 fats: Can we find a sustainable substitute for fish? *Nutrients* 5(4): 1301–1315.

Li, C., Li, Z., Fan, M., Cheng, W., Long, Y., Ding, T., and Ming, L. 2006. The composition of *Hirsutella sinensis*, anamorph of *Cordyceps sinensis*. *Journal of Food Composition and Analysis* 19(8): 800–805.

Li, Y. T., Li, M. T., Fu, C. H., Zhou, P. P., Liu, J. M., and Yu, L. J. 2009. Improvement of arachidonic acid and eicosapentaenoic acid production by increasing the copy number of the genes encoding fatty acid desaturase and elongase into *Pichia pastoris*. *Biotechnology Letters* 31(7): 1011–1017.

Liang, M. H., and Jiang, J. G. 2013. Advancing oleaginous microorganisms to produce lipid via metabolic engineering technology. *Progress in Lipid Research* 52(4): 395–408.

Liang, Y., Zhao, X., Strait, M., and Wen, Z. 2012. Use of dry-milling derived thin stillage for producing eicosapentaenoic acid (EPA) by the fungus *Pythium irregulare*. *Bioresource Technology* 111: 404–409.

Liu, H., Cheng, T., Xian, M., Cao, Y., Fang, F., and Zou, H. 2012. Fatty acid from the renewable sources: A promising feedstock for the production of biofuels and biobased chemicals. *Biotechnology Advances* 32(2): 382–389.

Liu, G., Wang, H., Zhou, B., Guo, X., and Hu, X. 2010. Compositional analysis and nutritional studies of *Tricholoma matsutake* collected from Southwest China. *Journal of Medicinal Plants Research* 4(12): 1222–1227.

Lowry, R. R., and Tinsley, I. J. 1966. Oleic and linoleic acid interaction in polyunsaturated fatty acid metabolism in the rat. *The Journal of Nutrition* 88(1): 26–32.

Mahaffey, K. R., Clickner, R. P., and Jeffries, R. A. 2008. Methylmercury and omega-3 fatty acids: Co-occurrence of dietary sources with emphasis on fish and shellfish. *Environmental Research* 107(1): 20–29.

Majesty, D., Ijeoma, E., Winner, K., and Prince, O. 2019. Nutritional, anti-nutritional and biochemical studies on the oyster mushroom, *Pleurotus ostreatus*. *EC Nutrition* 14(1): 36–59.

Marekov, I., Momchilova, S., Grung, B., and Nikolova-Damyanova, B. 2012. Fatty acid composition of wild mushroom species of order Agaricales-Examination by gas chromatography–mass spectrometry and chemometrics. *Journal of Chromatography: Part B* 910: 54–60. doi:10.1016/j.jchromb.2012.10.025.

Martins, M. B., Suaiden, A. S., Piotto, R. F., and Barbosa, M. 2008. Propriedades dos ácidos graxos poliinsaturados–Omega 3 obtidos de óleo de peixe e óleo de linhaça Properties of Omega-3 polyunsaturated fatty acids obtained of fish oil and flaxseed oil. *Revista do instituto de ciências da Saúde* 26(2): 153–156.

Maskrey, B. H., Megson, I. L., Whitfield, P. D., and Rossi, A. G. 2011. Mechanisms of resolution of inflammation: A focus on cardiovascular disease. *Arteriosclerosis, Thrombosis, and Vascular Biology* 31(5): 1001–1006.

Mataix, J., Mañas, M., Llopis, J., and Martínez de Vitoria, E. 2003. *Tabla de composición de alimentos españoles*. 4th ed. Granada: Universidad de Granada, Instituto de Nutricion y Tecnologia de los Alimentos.

Meluzzi, A., Sirri, F., Manfreda, G., Tallarico, N., and Franchini, A. 2000. Effects of dietary vitamin E on the quality of table eggs enriched with n-3 long-chain fatty acids. *Poultry Science* 79(4): 539–545.

Meyer, B. J., Mann, N. J., Lewis, J. L., Milligan, G. C., Sinclair, A. J., and Howe, P. R. 2003. Dietary intakes and food sources of omega-6 and omega-3 polyunsaturated fatty acids. *Lipids* 38(4): 391–398.

Nakalembe, I., and Kabasa, J. D. 2013. Fatty and amino acids composition of selected wild edible mushrooms of Bunyoro sub-region, Uganda. *African Journal of Food, Agriculture, Nutrition and Development* 13(1): 7225–7241.

Napier, J. A., and Graham, I. A. 2010. Tailoring plant lipid composition: Designer oilseeds come of age. *Current Opinion in Plant Biology* 13(3): 329–336.

Obodai, M., Ferreira, I. C., Fernandes, Â., Barros, L., Mensah, D. L. N., Dzomeku, M., Urben, A. F., Prempeh, J., and Takli, R. K. 2014. Evaluation of the chemical and antioxidant properties of wild and cultivated mushrooms of Ghana. *Molecules* 19(12): 19532–19548.

Oliveira, P., Calado, M. L., and Rosado, J. 2003. Etnomicologia: uma breve introdução. In: A. S. Dias (Ed.), *Etnobotânica. Perspectivas, história e utilizações* (pp. 39–44). Évora: Universidade de Évora, Série Ciências da Natureza e do Ambiente.

Öztürk, M., Duru, M. E., Kivrak, Ş., Mercan-Doğan, N., Türkoglu, A., and Özler, M. A. 2011. *In vitro* antioxidant, anticholinesterase and antimicrobial activity studies on three *Agaricus* species with fatty acid compositions and iron contents: A comparative study on the three most edible mushrooms. *Food and Chemical Toxicology: An International Journal Published for the British Industrial Biological Research Association* 49(6): 1353–1360.

Pacetti, D., Mozzon, M., Lucci, P., and Frega, G. 2013. Bioactive fish fatty acids: Health effects and their use as functional food ingredients. *Current Nutrition and Food Science* 9(4): 283–297.

Pahl, S. L., Lewis, D. M., Chen, F., and King, K. D. 2010. Heterotrophic growth and nutritional aspects of the diatom *Cyclotella cryptica* (*Bacillariophyceae*): Effect of some environmental factors. *Journal of Bioscience and Bioengineering* 109(3): 235–239.

Pal, D., Khozin-Goldberg, I., Cohen, Z., and Boussiba, S. 2011. The effect of light, salinity, and nitrogen availability on lipid production by *Nannochloropsis* sp. *Applied Microbiology and Biotechnology* 90(4): 1429–1441.

Patil, V., Källqvist, T., Olsen, E., Vogt, G., and Gislerød, H. R. 2007. Fatty acid composition of 12 microalgae for possible use in aquaculture feed. *Aquaculture International* 15(1): 1–9.

Pedneault, K., Angers, P., Gosselin, A., and Tweddell, R. J. 2006. Fatty acid composition of lipids from mushrooms belonging to the family *Boletaceae*. *Mycological Research* 110(10): 1179–1183.

Pedneault, K., Angers, P., Gosselin, A., and Tweddell, R. J. 2008. Fatty acid profiles of polar and neutral lipids of ten species of higher Basidiomycetes indigenous to eastern Canada. *Mycological Research* 112, 1428–1434.

Pereira, E., Barros, L., Martins, A., and Ferreira, I. C. 2012. Towards chemical and nutritional inventory of Portuguese wild edible mushrooms in different habitats. *Food Chemistry* 130(2): 394–403.

Petrie, J. R., Shrestha, P., Belide, S., Kennedy, Y., Lester, G., Liu, Q., Divi, U. K., Mulder, R. J., Mansour, M. P., Nichols, P. D., and Singh, S. P. 2014. Metabolic engineering *Camelina sativa* with fish oil-like levels of DHA. *PLoS One* 9(1): e85061.

Petrie, J. R., Shrestha, P., Mansour, M. P., Nichols, P. D., Liu, Q., and Singh, S. P. 2010. Metabolic engineering of omega-3 long-chain polyunsaturated fatty acids in plants using an acyl-CoA Δ6-desaturase with ω3-preference from the marine microalga *Micromonas pusilla*. *Metabolic Engineering* 12(3): 233–240.

Petrović, J., Stojković, D., Reis, F. S., Barros, L., Glamočlija, J., Ćirić, A., Ferreira, I. C., and Soković, M. 2014. Study on chemical, bioactive and food preserving properties of *Laetiporus sulphureus* (Bull.: Fr.) Murr. *Food and Function* 5(7): 1441–1451.

Phinney, S. D., Odin, R. S., Johnson, S. B., and Holman, R. T. 1990. Reduced arachidonate in serum phospholipids and cholesteryl esters associated with vegetarian diets in humans. *The American Journal of Clinical Nutrition* 51(3): 385–392.

Punia, S., Sandhu, K. S., Siroha, A. K., and Dhull, S. B. 2019. Omega 3: Metabolism, absorption, bioavailability and health benefits: A review. *Pharma Nutrition* 10: 100162.

Puri, M. 2017. Algal biotechnology for pursuing omega-3 fatty acid (bioactive) production. *Microbiology Australia* 38(2): 85–88.

Racine, R. A., and Deckelbaum, R. J. 2007. Sources of the very-long-chain unsaturated omega-3 fatty acids: Eicosapentaenoic acid and docosahexaenoic acid. *Current Opinion in Clinical Nutrition and Metabolic Care* 10(2): 123–128.

Raper, N. R., Cronin, F. J., and Exler, J. 1992. Omega-3 fatty acid content of the US food supply. *Journal of the American College of Nutrition* 11(3): 304–308.

Ribeiro, B., de Pinho, P. G., Andrade, P. B., Baptista, P., and Valentão, P. 2009. Fatty acid composition of wild edible mushrooms species: A comparative study. *Microchemical Journal* 93(1): 29–35.

Rogalski, M., and Carrer, H. 2011. Engineering plastid fatty acid biosynthesis to improve food quality and biofuel production in higher plants. *Plant Biotechnology Journal* 9(5): 554–564.

Rubio-Rodríguez, N., Beltrán, S., Jaime, I., Sara, M., Sanz, M. T., and Carballido, J. R. 2010. Production of omega-3 polyunsaturated fatty acid concentrates: A review. *Innovative Food Science and Emerging Technologies* 11(1): 1–12.

Ruiz-Lopez, N., Haslam, R. P., Napier, J. A., and Sayanova, O. 2014. Successful high-level accumulation of fish oil omega-3 long-chain polyunsaturated fatty acids in a transgenic oilseed crop. *The Plant Journal* 77(2): 198–208.

Ruiz-López, N., Sayanova, O., Napier, J. A., and Haslam, R. P. 2012. Metabolic engineering of the omega-3 long chain polyunsaturated fatty acid biosynthetic pathway into transgenic plants. *Journal of Experimental Botany* 63(7): 2397–2410.

Ryan, J., Farr, H., Visnovsky, S., Vyssotski, M., and Visnovsky, G. 2010. A rapid method for the isolation of eicosapentaenoic acid-producing marine bacteria. *Journal of Microbiological Methods* 82(1): 49–53.

Ryckebosch, E., Muylaert, K., and Foubert, I. 2012. Optimization of an analytical procedure for extraction of lipids from microalgae. *Journal of the American Oil Chemists' Society* 89(2): 189–198.

Sadiq, S., Bhatti, H. N., and Hanif, M. A. 2008. Studies on chemical composition and nutritive evaluation of wild edible mushrooms. *Iranian Journal of Chemistry and Chemical Engineering* 27(3): 151–154.

Sakuradani, E., Ando, A., Shimizu, S., and Ogawa, J. 2013. Metabolic engineering for the production of polyunsaturated fatty acids by oleaginous fungus *Mortierella alpina* 1S-4. *Journal of Bioscience and Bioengineering* 116(4): 417–422.

Salunke, D., Mangalekar, R., and Kuvalekar, A. 2014. Bioconversion of alpha-linolenic acid into long chain polyunsaturated fatty acids by oleaginous fungi. *International Journal of Pharmacy and Biological Sciences* 5(2): 27–35.

Sande, D., Colen, G., dos Santos, G. F., Ferraz, V. P., and Takahashi, J. A. 2019a. Production of omega 3, 6, and 9 fatty acids from hydrolysis of vegetable oils and animal fat with *Colletotrichum gloeosporioides* lipase. *Food Science and Biotechnology* 27(2): 537–545.

Sande, D., de Oliveira, G. P., Moura, M. A. F., de Almeida Martins, B., Lima, M. T. N. S., and Takahashi, J. A. 2019b. Edible mushrooms as a ubiquitous source of essential fatty acids. *Food Research International* 125: 108524.

Sang, M., Wang, M., Liu, J., Zhang, C., and Li, A. 2012. Effects of temperature, salinity, light intensity, and pH on the eicosapentaenoic acid production of *Pinguiococcus pyrenoidosus*. *Journal of Ocean University of China* 11(2): 181–186.

Santos-Merino, M., Garcillán-Barcia, M. P., and de la Cruz, F. 2018. Engineering the fatty acid synthesis pathway in *Synechococcus elongatus* PCC 7942 improves omega-3 fatty acid production. *Biotechnology for Biofuels* 11(1): 239.

Sayanova, O., and Napier, J. A. 2011. Transgenic oilseed crops as an alternative to fish oils. *Prostaglandins, Leukotrienes and Essential Fatty Acids (PLEFA)* 85(5): 253–260.

Schenk, P. M., Thomas-Hall, S. R., Stephens, E., Marx, U. C., Mussgnug, J. H., Posten, C., Kruse, O., and Hankamer, B. 2008. Second generation biofuels: High-efficiency microalgae for biodiesel production. *Bioenergy Research* 1(1): 20–43.

Scott, S. D., Armenta, R. E., Berryman, K. T., and Norman, A. W. 2011. Use of raw glycerol to produce oil rich in polyunsaturated fatty acids by a thraustochytrid. *Enzyme and Microbial Technology* 48(3): 267–272.

Shao, S., Hernandez, M., Kramer, J. K. G., Rinker, D. L., and Tsao, R. 2010. Ergosterol profiles, fatty acid composition, and antioxidant activities of button mushrooms as affected by tissue part and developmental stage. *Journal of Agricultural and Food Chemistry* 58(22): 11616–11625.

Shapira, N., Weill, P., and Loewenbach, R. 2008. Egg fortification with n-3 polyunsaturated fatty acids (PUFA): Nutritional benefits versus high n-6 PUFA western diets, and consumer acceptance. *The Israel Medical Association Journal: IMAJ* 10(4): 262–265.

Shoeb, M., Mondal, P., Kylin, H., and Nahar, N. 2017. Chemical and biological activity of mushrooms naturally occurring in Bangladesh. *Asian Journal of Pharmacognosy* 1(3): 42–51.

Sijtsma, L., and De Swaaf, M. E. 2004. Biotechnological production and applications of the ω-3 polyunsaturated fatty acid docosahexaenoic acid. *Applied Microbiology and Biotechnology* 64(2): 146–153.

Simopoulos, A. P. 2002a. Genetic variation and dietary response: nutrigenetics/nutrigenomics. *Asia Pacific Journal of Clinical Nutrition* 11: S117–S128.

Simopoulos, A. P. 2002b. Omega-3 fatty acids in inflammation and autoimmune diseases. *Journal of the American College of Nutrition* 21(6): 495–505.

Steinrücken, P., Erga, S. R., Mjøs, S. A., Kleivdal, H., and Prestegard, S. K. 2017. Bioprospecting North Atlantic microalgae with fast growth and high polyunsaturated fatty acid (PUFA) content for microalgae-based technologies. *Algal Research* 26: 392–401.

Suman, K., Kiran, T., Devi, U. K., and Sarma, N. S. 2012. Culture medium optimization and lipid profiling of *Cylindrotheca*, a lipid-and polyunsaturated fatty acid-rich pennate diatom and potential source of eicosapentaenoic acid. *Botanica Marina* 55(3): 289–299.

Swanson, D., Block, R., and Mousa, S. A. 2012. Omega-3 fatty acids EPA and DHA: Health benefits throughout life. *Advances in Nutrition* 3(1): 1–7.

Swinbanks, D. 1993. Fish-oil miracle additive brings benefits to some. *Nature* 364(6434): 180–180.

Tiez, L., and Zeiger, E. 2010. *Plant Physiology*. 5th edition, 1–778. Sinauer Associates Inc, Publishers: Sunderland.

Tocher, D. R., Betancor, M. B., Sprague, M., Olsen, R. E., and Napier, J. A. 2019. Omega-3 long-chain polyunsaturated fatty acids, EPA and DHA: Bridging the gap between supply and demand. *Nutrients* 11(1): 89.

Vadivelan, G., and Venkateswaran, G. 2014. Production and enhancement of omega-3 fatty acid from *Mortierella alpina* CFR-GV15: Its food and therapeutic application. *BioMed Research International* 9. doi:10.1155/2014/657414.

Valverde, M. E., Hernández-Pérez, T., and Paredes-López, O. 2015. Edible mushrooms: Improving human health and promoting quality life. *International Journal of Microbiology* 14. doi:10.1155/2015/376387.

Van Wagenen, J., Miller, T. W., Hobbs, S., Hook, P., Crowe, B., and Huesemann, M. 2012. Effects of light and temperature on fatty acid production in *Nannochloropsis salina*. *Energies* 5(3): 731–740.

Vargas-Isla, R., Ishikawa, N. K., and Py-Daniel, V. 2013. Contribuições etnomicológicas dos povos indígenas da Amazônia. *Biota Amazônia (Biote Amazonie, Biota Amazonia, Amazonian Biota)* 3(1): 58–65.

Verma, M. K., Halade, G. V., and Black, L. M. 2018. Paradigm shift: Metabolic transformation of docosahexaenoic and eicosapentaenoic acids to bioactives exemplify the promise of fatty acid drug discovery. *Biotechnology Advances* 36(4): 935–953.

Verma, M. L., Kishor, K., Sharma, D., Kumar, S., and Sharma, K. D. 2020. Microbial production of omega-3 polyunsaturated fatty acids. In: Verma, M. L. and Chandel, A. K. (eds), *Biotechnological Production of Bioactive Compounds*, 293–326. Elsevier: Amsterdam.

Ward, O. P., and Singh, A. 2005. Omega-3/6 fatty acids: Alternative sources of production. *Process Biochemistry* 40(12): 3627–3652.

Wen, Z. Y., and Chen, F. 2001. Optimization of nitrogen sources for heterotrophic production of eicosapentaenoic acid by the diatom *Nitzschia laevis*. *Enzyme and Microbial Technology* 29(6–7): 341–347.

Wen, Z. Y., and Chen, F. 2003. Heterotrophic production of eicosapentaenoic acid by microalgae. *Biotechnology Advances* 21(4): 273–294.

Wijesundera, C., Ceccato, C., Watkins, P., Fagan, P., Fraser, B., Thienthong, N., and Perlmutter, P. 2008. Docosahexaenoic acid is more stable to oxidation when located at the sn-2 position of triacylglycerol compared to sn-1 (3). *Journal of the American Oil Chemists' Society* 85(6): 543–548.

Winwood, R. J. 2013. Recent developments in the commercial production of DHA and EPA rich oils from micro-algae. *Oilseeds and Fats, Crops and Lipids* 20(6): 604.

Woldegiorgis, A. Z., Abate, D., Haki, G. D., Ziegler, G. R., and Harvatine, K. J. 2015. Fatty acid profile of wild and cultivated edible mushrooms collected from Ethiopia. *Journal of Nutrition and Food Sciences* 5(3): 1.

Wu, G., Truksa, M., Datla, N., Vrinten, P., Bauer, J., Zank, T., Cirpus, P., Heinz, E., and Qiu, X. 2005a. Stepwise engineering to produce high yields of very long-chain polyunsaturated fatty acids in plants. *Nature Biotechnology* 23(8): 1013–1017.

Wu, S. T., Yu, S. T., and Lin, L. P. 2005b. Effect of culture conditions on docosahexaenoic acid production by *Schizochytrium* sp. S31. *Process Biochemistry* 40(9): 3103–3108.

Xiao-Hong, C. A. O., Song-Yao, L. I., Wang, C. L., and Mei-Fang, L. U. 2007. Potential use of the herbicide quizalofop-p-ethyl for eicosapentaenoic acid overproduction by the diatom *Nitzschia laevis*. *Sheng Wu Gong Cheng Xue Bao = Chinese Journal of Biotechnology* 23(5): 885–890.

Xie, D., Jackson, E. N., and Zhu, Q. 2015. Sustainable source of omega-3 eicosapentaenoic acid from metabolically engineered *Yarrowia lipolytica*: From fundamental research to commercial production. *Applied Microbiology and Biotechnology* 99(4): 1599–1610.

Xue, Z., Sharpe, P. L., Hong, S. P., Yadav, N. S., Xie, D., Short, D. R., Damude, H. G., Rupert, R. A., Seip, J. E., Wang, J., Pollak, D. W., Bostick, M. W., Bosak, M. D., Macool, D. J., Hollerbach, D. H., Zhang, H., Arcilla, D. M., Bledsoe, S. A.,

Croker, K., McCord, E. F., Tyreus, B. D., Jackson, E. N., and Pollak, D. W. 2013. Production of omega-3 eicosapentaenoic acid by metabolic engineering of *Yarrowia lipolytica*. *Nature Biotechnology* 31(8): 734.

Yazawa, K. 1996. Production of eicosapentaenoic acid from marine bacteria. *Lipids* 31(1Part2): 297–300.

Yazawa, K., Araki, K., Okazaki, N., Watanabe, K., Ishikawa, C., Inouc, A., Numao, N., and Kondo, K. 1988. Production of eicosapentaenoic acid by marine bacteria. *The Journal of Biochemistry* 103(1): 5–7.

Yilmaz, N., Solmaz, M., Türkekul, I., and Elmastas, M. 2006. Fatty acid composition in some wild edible mushrooms growing in the middle Black Sea region of Turkey. *Food Chemistry* 99(1): 168–174.

Yongmanitchai, W. A. R. D., and Ward, O. P. 1991. Growth of and omega-3 fatty acid production by *Phaeodactylum tricornutum* under different culture conditions. *Applied and Environmental Microbiology* 57(2): 419–425.

Zahringer, U., Domergue, F., Abbadi, A., Moreau, H. X. E., and Heinz, E. 2005. *In vivo* characterization of the first acyl-CoA Delta6-desaturase from a member of the plant kingdom, the microalga *Ostreococcus tauri*. *Biochemistry Journal* 389: 483–490.

Zárate, R., el Jaber-Vazdekis, N., and Ramírez-Moreno, R. 2016. Importance of polyunsaturated fatty acids from marine algae. In: Hegde, M., Zanwar, A., and Adekar, S. (eds)., *Omega-3 Fatty Acids*, 101–126. Springer: Cham.

Zhu, L., Zhang, X., Ji, L., Song, X., and Kuang, C. 2007. Changes of lipid content and fatty acid composition of *Schizochytrium limacinum* in response to different temperatures and salinities. *Process Biochemistry* 42(2): 210–214.

3

Structural Chemistry of Essential Fatty Acids

Prince Chawla, Mudasir Yaqoob, Aarti Bains, Rohit Chhetri and Ravinder Kaushik

Contents

3.1 INTRODUCTION

Essential fatty acids (EFAs) are well established as potent components of the human diet that cannot be synthesized in vivo; however, essential fatty acids are inevitably required for maintenance of optimal human health (Mogensen, 2017; Punia et al., 2019). In addition, essential fatty acids are the vital component of dietary fat and several dietary sources are required to obtain the essential fatty acids (Virsangbhai et al., 2019). Apart from normal metabolism and proper human health, essential fatty acids are required by the body to produce crucial cell membrane constituents, and in terms of membranes' properties of fluidity, to determine and affect the behavior of enzymes and receptors bound with the cell membrane (Calder, 2015; De Carvalho and Caramujo, 2018; Amiri et al., 2018). Moreover, it is well known that the term "fatty acids" always refers to polyunsaturated fatty acids (PUFA); there are two families of essential fatty acids, i.e., ω-3 (omega-3) and ω-6 (omega-6) (Virsangbhai et al., 2019). The omega-3 fatty acid family consists mostly of linolenic acid, docosahexaenoic acid and docosapentaenoic acid, whereas linoleic acid and arachidonic acid are classified as omega-6 fatty acids. Basically, various food sources such as flax seeds, hemp, walnut, soybean, canola oil and chloroplasts of green leafy vegetables are the efficient sources of α-linolenic acid (Hunter, 1990; Kaur et al., 2014). On the other hand, eicosapentaenoic acid is a major component of eicosanoid hormones and it can be obtained from both fresh- and saltwater fish and fish oil (Kinsella, 1990; Kaur et al., 2014). As well, docosahexaenoic acid is principally present in red-brown algae and fish oil and it is considered as a major brain (50% DHA) fatty acid (Kaur et al., 2014). Safflower, sunflower, meat, eggs, dairy products, soybean, sesame, almonds and corn are the major sources of omega-6 fatty acids (Horrobin, 1990; Cholewski et al., 2018). The essential fatty acids that are agglomerated into phospholipids are mainly important in the whole structure and function of both omega-6 and omega-3 fatty acids, as phospholipids sustain both the structural integrity and the critical functioning of cellular membranes throughout the body (Surette, 2008; Kaur et al., 2014; Cholewski et al., 2018). Herein, omega-3 fatty acids consist of a final carbon–carbon double bond in the third bond of the fatty acid–containing methyl end (i.e., ω-3 position), whereas omega-6 fatty acids consist of a final carbon–carbon double bond in the sixth bond of the fatty acid–containing methyl end of the fatty acid (i.e., ω-6 position) as described in Figure 3.1 (a & b) (Kaur et al., 2014; Cholewski et al., 2018).

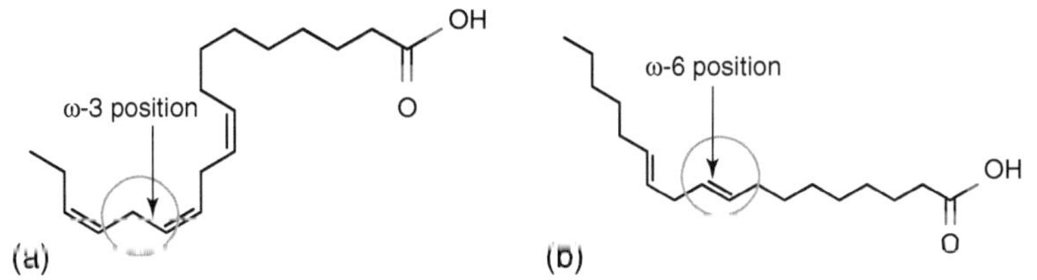

Figure 3.1 Basic structure of (a) omega-3 and (b) omega-6 fatty acids

The nomenclature of fatty acids is an important feature, and in the case of essential fatty acids, systematic nomenclature and common names are used to describe the essential fatty acids. The number and position of double bonds indicate the *cis* and *trans* forms of the essential fatty acids (Vannice and Rasmussen, 2014; Cholewski et al., 2018). Alike, in saturated fatty acids, the numerical abbreviation system is also used and the second number confirms the presence of several double bonds (Ratnayake and Galli, 2009). Therefore, in this chapter the structural chemistry is described in detail for a better understanding of the essential fatty acids. Nomenclatures of fatty acids are also discussed in detail with suitable examples for better readability and understanding.

3.2 NOMENCLATURE OF ESSENTIAL FATTY ACIDS

Essential fatty acids consist of double bonds in their aliphatic chain system; according to the International Union of Pure and Applied Chemistry (IUPAC) system, term the *anoic* is transformed into the *enoic* to enable the occurrence of a double bond. As well, with the number of double bonds in essential fatty acids, the terms *di, tri, tetra* and so on are also added, and with the exception of some of the long-chain polyunsaturated fatty acids, common names of essential fatty acids are also in use. Similar to saturated fatty acids, for essential fatty acids, the numerical abbreviation system is also used, with the second number signifying the presence of several double bonds. Also, to indicate the positions of double bonds of the essential fatty acids, the Δ system is used, with numbering starting at the carboxylic acid carbon. To understand the concept, take the example of a linoleic acid that has 18 carbons with 2 double bonds: this could be considered as 9,12-octadecadienoic acid. Furthermore, a numbering system that represents the position of double bonds from the methyl terminal end of the fatty acids is known as the omega system of the essential fatty acids. Based upon the biological activity and biosynthetic origin, the omega system is so often useful to bring together the fatty acids in a group since several enzymes distinguish fatty acids from the free methyl end of the molecule after it is esterified to glycerol. Also, the natural configuration or pattern of double bonds in an essential fatty acid is the *cis* configuration, in which carbons of the aliphatic chain lengthen on the same side as the double bonds. Conversely, the *trans'* form of essential fatty acids consists of carbon atoms on the opposite side. Furthermore, there is the pentadiene system, where double bonds of essential fatty acids are usually methylene interrupted, or in this system, the 2 double bonds situated at carbon 1 and carbon 4 (double bonds are separated by the carbon of the methylene terminal end). Therefore, it is probable to foresee the exact position of all the double bonds in most essential fatty acids if the site of the first double bond along with the aliphatic carbon chain is known. All the essential fatty acids reveal varied melting points, and the presence of double bonds in essential fatty acids affects the melting points of the fatty acids. Essential fatty acids do not exhibit linear structures or configurations due to bent configuration resulting from the presence of the double bonds. Due to this nature, essential fatty acids are not able to orient themselves into tight packing conformations and Van der Waals interactions between the molecules of essential fatty acids are comparatively weak. To endorse the solid-liquid transition phase, due to weak forces, less energy is required, which results in decreased melting points for essential fatty acids.

3.3 CHEMICAL STRUCTURES OF OMEGA-3 AND OMEGA-6 FATTY ACIDS

3.3.1 Omega-3 Fatty Acids

3.3.1.1 Linolenic Acid

Linolenic acid belongs to the group of omega-3 fatty acids and is mostly present in the fish oils. This type of fatty acid has been characterized into two types based upon differing structures: alpha-linolenic (ALA) and gamma-linolenic (GLA) acids. As well, both types of essential fatty acids contain 18 carbon atoms along with three double bonds in *cis* form. Moreover, ALA is a form of carboxylic acid that consists of 18 carbon atoms with three double bonds. According to the IUPAC system, the systematic nomenclature is all-*cis*-9,12,15-octadecatrienoic acid; herein, all-*cis* 9,12,15 depicts the position of the *cis* double bonds on the carbon atoms from the functional group (Aldred, 2008). The first double bond is located on the 15th carbon atom from the functional group (towards the terminal carbon methyl group). It is also known as the isomer of gamma-linolenic acid. The chemical formula is $C_{18}H_{30}O_2$ and it has a molar mass of 278.436 g/mol. The density of this compound is 0.9164 g/cm^3 (Figure 3.2). Linolenic acid is essential for the production of docosahexaenoic acid and eicosapentaenoic acid, and in the same way linolenic acid acts as the precursor of leukotrienes, prostaglandins and thromboxans, all eicosanoids that are important for the modulation of certain diseases like thrombosis and inflammation (Bames and Tappenden, 2017).

3.3.1.2 Docosahexaenoic Acid

Docosahexaenoic acid (DHA) is the chief structural component of skin, retina, brain and cerebral cortex (Guesnet and Alessandri, 2011). It belongs to the family of omega-3 fatty acids and is represented with the formula $C_{22}H_{32}O_2$. The structure consists of an open-chain structure of 22 carbons, attached with a functional carboxylic group possessing six double bonds, which are represented as 22:6; wherein 22 represents the number of carbon atoms and 6 is the number of double bonds (n=6). All the double bonds are in *cis* from. The 1st double bond is located at the 3rd carbon from the methyl end that is why the fatty acid falls in the category of omega-3 fatty acids. The positions of the double bonds (as depicted in Figure 3.3) are

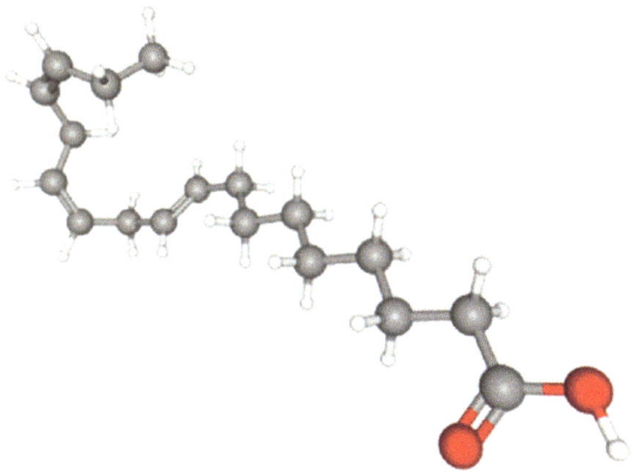

Figure 3.2 Chemical representation of alpha-linolenic acid

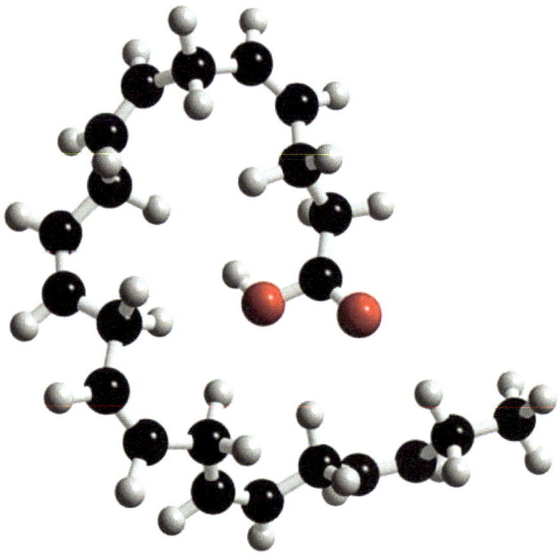

Figure 3.3 Schematic representation of docosahexaenoic acid

on carbon numbers 4, 7, 10, 13, 16 and 19. The common name of the DHA is cervonic acid and the IUPAC name is all-*cis*-docosa-4,7,10,13,16,19-hexa-enoic acid 22:6 (n=6). This polyunsaturated fatty acid is important for improving the brain development of newborn babies. The recommended intake of DHA for lactating and pregnant women is 200–300 mg/day.

3.3.1.3 Docosapentaenoic Acid

Docosapentaenoic acid (DPA) belongs to the group of polyunsaturated fatty acids, which are mainly formed as an intermediate fatty acid species between eicosapentaenoic acid and docosahexaenoic acid (Richter et al., 2016). The essential fatty acid belongs to the omega-3 fatty acids and contains 22 carbon atoms with 5 double bonds in *cis* isomeric form. The *cis* form means that the hydrogen atoms adjacent to carbon at the double bonds protrude to the same side of the chain (Figure 3.4). The 1st double bond is positioned at carbon number 3 from the methyl terminal end. DPA is chiefly used to differentiate two isomers, all-*cis*-7,10,13,16,19-docosapentaenoic acid and all-*cis*-4,7,10,13,16-docosapentaenoic acid, which are usually designated as n-3 DPA and n-6 DPA, which are isomers of each other. The numbers 7, 10, 13, 16 and 19 depict the positions of the double bonds from the functional group. The two designations of

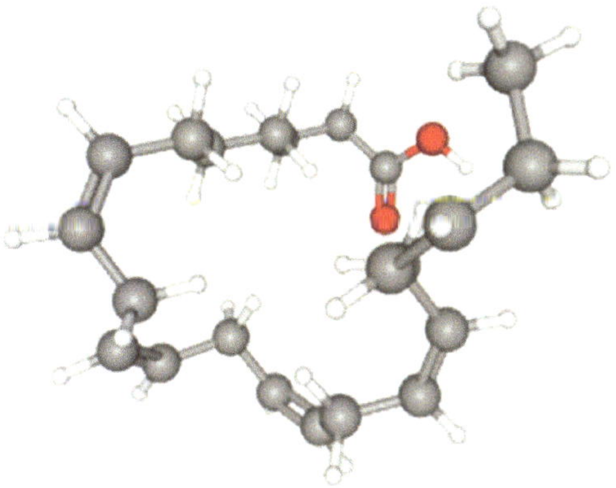

Figure 3.4 Schematic representation of docosapentaenoic acid

DHA usually differentiate the two classes of fatty acids as ω-3 and ω-6 (the basis of the position of the double bonds from the methyl terminal end). The common names of the two DHA are clupanodonic acid and bond acid. The molecular formula is $C_{22}H_{34}O_2$ and it has a molecular weight of 330.5 g/mol. Clupanodonic acid is associated with lowering inflammation (Calder et al., 2013).

3.3.1.4 Eicosapentaenoic Acid

Eicosapentaenoic acid (EPA) belongs to the group of omega-3 fatty acids that are synthesized from ALA in the human body (Pillsbury et al., 2011). EPA is sometimes called the breakdown product of eicosapentaenoyl ethanolamide (Lucanic et al., 2011). EPA is a 20-carbon chain with a functional group of a carboxylic acid–containing 5-double bond with the chemical formula $C_{20}H_{30}O_2$. The location of the 1st double bond is at carbon number 3 from the omega side. It is usually represented as 20:5 (n-3); "20" depicting the number of carbons; "5" as the number of double bonds and "n-3" the omega-3 fatty acid group (Figure 3.5). The IUPAC name of eicosapentaenoic acid is (5Z,8Z,11Z,14Z,17Z)-5,8,11,14,17-eicosapentaenoic acid and the trivial name is timnodonic acid. The molar mass of this compound is 302.451 g/mol with a melting point of –54 °C. The fatty acid has been reported to be responsible for preventing heart diseases (Mohebi-Nejad and Bikdeli, 2014). This PUFA is the precursor of leukotriene-5 eicosanoids, prostaglandin-3 and thromboxane-3, which have the capacity to inhibit platelet aggregation (Lucanic et al., 2011).

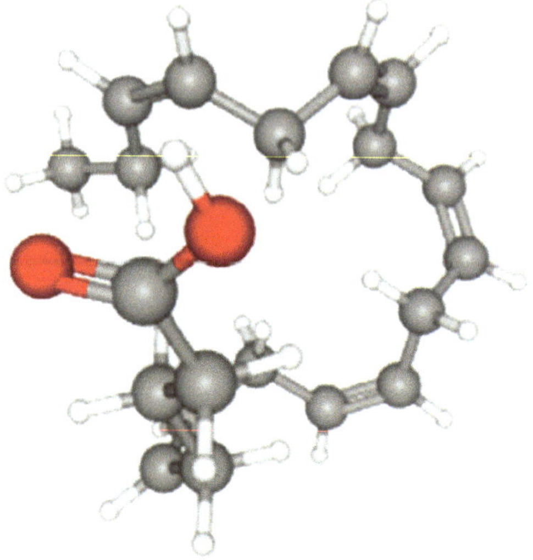

Figure 3.5 Schematic representation of eicosapentaenoic acid

3.3.2 Omega-6 Fatty Acids

Omega-6 fatty acids are the essential polyunsaturated fatty acids found in nuts, seeds and some vegetable oils. The members of this group can have been found to have anti-inflammatory properties (Nowak, 2010). Some examples of omega-6 fatty acids are the following.

3.3.2.1 Linoleic Acid

Linoleic acid is a polyunsaturated fatty acid made of an open chain of 18-carbon atoms and two double bonds. Omega-6 refers to the position of the double bond from the methyl terminal end. The positions of the two double bonds from the functional group are 9 and 12, all in *cis* configurations (Figure 3.6). The systemic name is (9Z,12Z)-octadeca-9,12-dienoic acid, where *octadeca* refers to the number of carbon atoms, and *9,12* to the position of the double bonds on the carbon atoms and *enoic acid* to a functional group attached. Sometimes in the literature, it is designated in a shorthand notation as 18:2 *cis*-9,12 (n-2). In nature, it usually occurs as triglyceride ester: free fatty acids which are the form that is not combined with glycerol to form triglyceride and are naturally low in foods (Kaplan et al., 1997). It is soluble in ethanol, benzene,

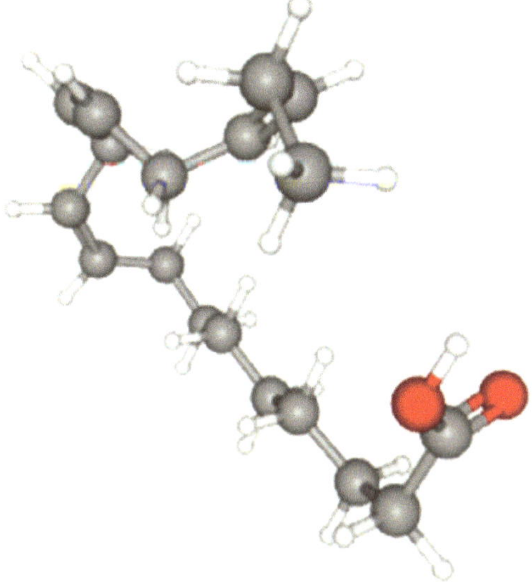

Figure 3.6 Schematic representation of linoleic acid

acetone and diethyl ether (William, 2016). Linoleic acid appears as a colorless to pale yellow liquid. Linoleic acid is designated with the molecular formula $C_{18}H_{32}O_2$. It has been found closely associated with plant glycosides. The melting and boiling points of this compound are –5 °C and 230 °C, respectively. Arachidonic acid is produced from linoleic acid (Sanders, 2016).

3.3.2.2 Arachidonic Acid

Arachidonic acid is a carboxylic acid containing 20 carbon atoms and 4 double bonds. It belongs to the family of omega-6 fatty acids represented as 20:4 (5, 8, 11, 14) or 20:4 (ω-6) (Figure 3.7). The double bond is also located at the 6 carbon from the omega end. The systemic name is (5Z,8Z,11Z,14Z)-5,8,11,14-eicosatetraenoic acid, where 5,8,11,14 describes the position of the double bonds on the carbon atoms. Structurally, it is correlated to arachidic acid, which is present in cupuaçu butter. The chemical formula is $C_{20}H_{32}O_2$ and it has a molecular mass of 304.46 g/mol. The melting point is –49 °C, the boiling point is 169 °C, and it is soluble in ethanol. Arachidonic acid is only synthesized from linoleic acid in the body. Arachidonic acid

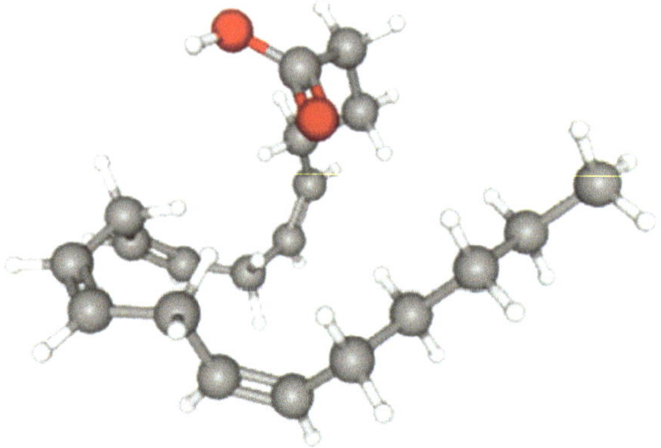

Figure 3.7 Schematic representation of arachidonic acid

is not considered as an essential fatty acid but becomes essential during a deficiency of linoleic acid.

3.3.2.3 Omega-9 Fatty Acids

Omega-9 fatty acids belong to the group of monounsaturated fatty acids, which are mostly found in animal and plant fats. The fatty acid is called omega-9 due to the location of the C-C double bond on the 9th position from the omega end. The fatty acid is represented as n-9 or 18:1; 18 is the number of carbon atoms and 1 is the number of the double bond (Bourre, 2004). There are generally two examples of omega-9 fatty acids which are most important in the food industry: erucic acid ((Z)-docos-13-enoic acid) and oleic acid ((Z)-octadec-9-enoic acid). Erucic acid comprises 22 carbon atoms, while oleic acid contains 18 carbon atoms with one double bond located between the 9th and 10th carbon atoms from the methyl end (Olson and Aryana, 2017). Omega-9 fatty acids are not considered completely essential as they can naturally be synthesized in the body from unsaturated fatty acids (Koutsos et al., 2016). Some common examples of omega-9 fatty acids are hypogeic acid (16:1 (n-9) (Z)-hexadec-7-enoic acid), elaidic acid (18:1 (n-9) (E)-octadec-9-enoic acid), gondoic acid (20:1 (n-9) (Z)-eicos-11-enoic acid), nervonic acid 24:1 (n-9) (Z)-tetracos-15-enoic acid and ximenic acid 26:1 (n-9). Figure 3.8 represents 3D structures of omega-9 fatty acids.

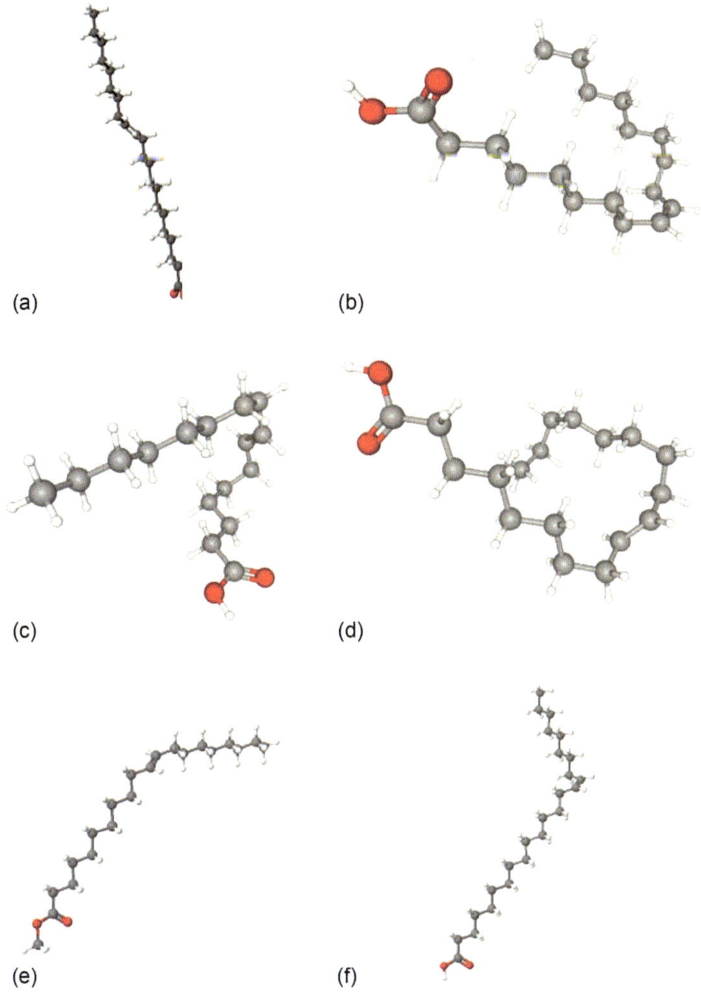

Figure 3.8 Schematic representation of omega-9 fatty acids: (a) erucic acid, (b) oleic acid, (c) hypogeic acid, (d) elaidic acid, (e) gondoic acid and (f) nervonic acid

3.4 STRUCTURAL MODIFICATION IN ESSENTIAL FATTY ACIDS

Omega-3 and omega-6 fatty acids are important in our diets because of their health benefits like their roles in cognition, neurodevelopment and neurodegenerative and heart disorders. Basically, EPA and DHA are usually synthesized from the dietary ALA by Sprecher's shunt: α-linolenic acid→stearidonic acid (SDA)→eicosatetraenoic acid (ETA)→eicosapentaenoic acid (EPA)→docosapentaenoic acid (DPA)→9,12,15,18,21-tetracosapentaenoic acid (TPA)→6,9,12,15,18,21-yetracosahexaenoic acid (YTA)→docosahexaenoic acid (DHA) (Burdge and Calder, 2005). The conversion is due to the presence of two enzymes, elongase and desaturase, which cause the elongation and desaturation of the functional groups, respectively. Docosahexaenoic acid is synthesized through a carbon-24 intermediate followed by the beta-oxidation in the cells (usually referred to as peroximonal oxidation). Eicosapentaenoic acid is elongated two times, synthesizing 24:5, omega-3 and accordingly desaturated to ω-3, 24:6; then, this is converted to docosahexaenoic acid 22:6, ω-3 through beta-oxidation (Figure 3.9). The conversion of ALA to DHA and EPA is less than 10% in the body (Goyens et al., 2005; Swanson et al., 2012).

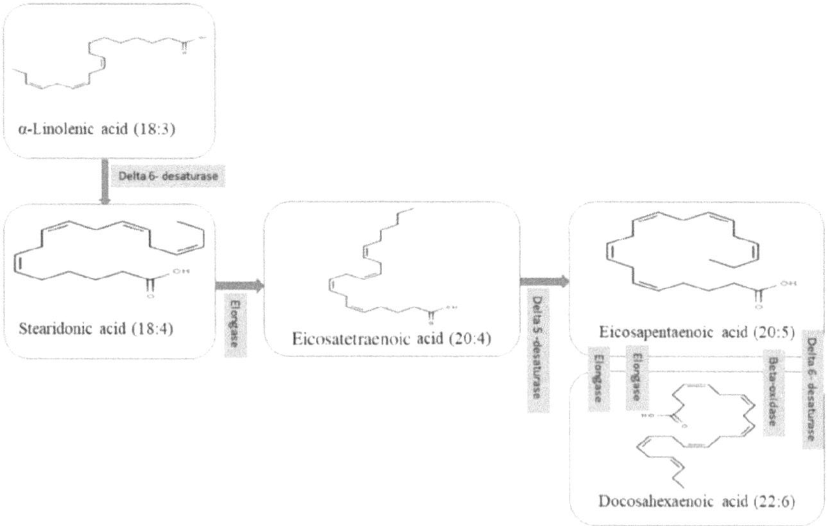

Figure 3.9 Schematic representation of structural modification in essential fatty acids

The steps involved for the bioconversion are as follows:

- Desaturation at 6th carbon (6-C) of ALA by Δ6 desaturase to yield SDA
- Elongation of SDA by Δ6 elongase, producing ETA
- Desaturation at 5th C of ETA by Δ5 desaturase, yielding EPA
- Elongation of EPA by Δ5 elongase to yield DPA and
- Desaturation at the fourth carbon of DPA by Δ4 desaturase to yield DHA (Simon et al., 2009).

3.5 CONCLUSION

The human body is not capable of synthesizing essential fatty acids; therefore, additional dietary sources are required to attain the recommended daily intake of the essential fatty acids. Owing to the fact that for normal metabolism and appropriate human health, essential fatty acids are crucially required to produce cell membrane components. The term "essential fatty acid" always designates a polyunsaturated fatty acid and includes two important categories: omega-3 and omega-6 fatty acids. Structurally, the number and position of double bonds in both types of essential fatty acids confirm the nomenclature of fatty acids. In the case of fatty acids, based upon their biological activity and biosynthetic origin, the omega system is more important and useful to bring together the fatty acids in a group, since the number of enzymes differentiates fatty acids from the free methyl end of the molecule after it is esterified to glycerol. In conclusion, the natural configuration or pattern of double bonds in an essential fatty acid is the *cis* configuration, in which carbons of the aliphatic chain lengthen on the same side as the double bonds. Conversely, the *trans'* form of essential fatty acids consists of carbon atoms on the opposite sides.

REFERENCES

Aldred, E. M. (2008). *Pharmacology E-Book: A Handbook for Complementary Healthcare Professionals.* UK: Elsevier Health Sciences.

Amiri, M., Yousefnia, S., Forootan, F. S., Peymani, M., Ghaedi, K., & Esfahani, M. H. N. (2018). Diverse roles of fatty acid binding proteins (FABPs) in development and pathogenesis of cancers. *Gene,* 676, 171–183.

Barnes, J. L., & Tappenden, K. A. (2017). Nutritional management of inflammatory bowel disease and short bowel syndrome. In: Coulston, A. M., Boushey, C. J., Ferruzi, M. G., & Delahanty, L.M. (Eds.), *Nutrition in the Prevention and Treatment of Disease* (pp. 857–874). Academic Press.

Bourre, J. M. (2004). Roles of unsaturated fatty acids (especially omega-3 fatty acids) in the brain at various ages and during ageing. *Journal of Nutrition, Health and Aging*, 8(3), 163–175.

Burdge, G. C., & Calder, P. C. (2005). Conversion of $\alpha $-linolenic acid to longer-chain polyunsaturated fatty acids in human adults. *Reproduction, Nutrition, Development*, 45(5), 581–597.

Calder, P. C. (2013). Omega-3 polyunsaturated fatty acids and inflammatory processes: Nutrition or pharmacology? *British Journal of Clinical Pharmacology*, 75(3), 645–662.

Calder, P. C. (2015). Functional roles of fatty acids and their effects on human health. *JPEN: Journal of Parenteral and Enteral Nutrition*, 39, 18S–32S.

Cholewski, M., Tomczykowa, M., & Tomczyk, M. (2018). A comprehensive review of chemistry, sources and bioavailability of omega-3 fatty acids. *Nutrients*, 10(11), 1662.

De Carvalho, C. C., & Caramujo, M. J. (2018). The various roles of fatty acids. *Molecules*, 23(10), 2583.

Goyens, P. L., Spilker, M. E., Zock, P. L., Katan, M. B., & Mensink, R. P. (2005). Compartmental modeling to quantify α-linolenic acid conversion after a longer-term intake of multiple tracer boluses. *Journal of Lipid Research*, 46(7), 1474–1483.

Guesnet, P., & Alessandri, J. M. (2011). Docosahexaenoic acid (DHA) and the developing central nervous system (CNS) Implications for dietary recommendations. *Biochimie*, 93(1), 7–12.

Haynes, William M. (2016). *CRC Handbook of Chemistry and Physics* (97th ed., pp. 3–338). Boca Raton: CRC Press. ISBN 978-1-4987-5429-3.

Horrobin, D. F. (1990). Gamma-Linolenic acid. *Rev Contemp Pharmac-other*, 1, 1–41.

Hunter, J. E. (1990). n-3 fatty acids from vegetable oils. *The American Journal of Clinical Nutrition*, 51(5), 809–814.

Kaplan, J. M., Siemers, W., & Grill, H. J. (1997). Effect of oral versus gastric delivery on gastric emptying of corn oil emulsions. *American Journal of Physiology: Regulatory, Integrative and Comparative Physiology*, 273(4), R1263–R1270.

Kaur, N., Chugh, V., & Gupta, A. K. (2014). Essential fatty acids as functional components of foods: A review. *Journal of Food Science and Technology*, 51(10), 2289–2303.

Kinsella, J. E. (1990). Sources of omega-3 fatty acids in human diets. In: Lees, R. S., & Karel, M. (Eds.) *Omega-3 Fatty Acids in Health and Disease*. New York: Marcel Dekker.

Koutsos, E., Gelis, S., & Echols, M. S. (2016). Advancements in nutrition and nutritional therapy. In: Speer, B. L. (Ed.), *Current Therapy in Avian Medicine and Surgery* (pp. 142–176). WB Saunders.

Lucanic, M., Held, J. M., Vantipalli, M. C., Klang, I. M., Graham, J. B., Gibson, B. W., & Gill, M. S. (2011). N-acylethanolamine signalling mediates the effect of diet on lifespan in *Caenorhabditis elegans*. *Nature*, 473(7346), 226.

Mogensen, K. M. (2017). Essential fatty acid deficiency. *Practical Gastroenterology*, 41(6), 37–44.

Mohebi-Nejad, A., & Bikdeli, B. (2014). Omega-3 supplements and cardiovascular diseases. *Tanaffos*, 13(1), 6.

Nowak, J. Z. (2010). Anti-inflammatory pro-resolving derivatives of omega-3 and omega-6 polyunsaturated fatty acids. *Postepy Higieny i Medycyny Doswiadczalnej (Online)*, 64, 115–132.

Olson, D., & Aryana, K. J. (2017). Omega-3 polyunsaturated fatty acids added to yogurt. In: Shah, N. P. (Ed.), *Yogurt in Health and Disease Prevention* (pp. 135–149). Academic Press.

Pillsbury, L., Oria, M., & Erdman, J. (Eds.) (2011). *Nutrition and Traumatic Brain Injury: Improving Acute and Subacute Health Outcomes in Military Personnel.* Japan: National Academies Press.

Punia, S., Sandhu, K. S., Siroha, A. K., & Dhull, S. B. (2019). Omega 3: Metabolism, absorption, bioavailability and health benefits: A review. *PharmaNutrition*, 100162.

Ratnayake, W. N., & Galli, C. (2009). Fat and fatty acid terminology, methods of analysis and fat digestion and metabolism: A background review paper. *Annals of Nutrition & Metabolism*, 55(1–3), 8.

Richter, C. K., Skulas-Ray, A. C., & Kris-Etherton, P. M. (2016). Recommended intake of fish and fish oils worldwide. In: Raatz, S. & Bibus, D. M. (Eds.), *Fish and Fish Oil in Health and Disease Prevention* (pp. 27–48). London: Academic Press.

Sanders, T. A. (2016). Introduction: The role of fats in human diet. In: *Functional Dietary Lipids* (pp. 1–20). Cambridge: Woodhead Publishing.

Simon, J. A., Chen, Y. H., & Bent, S. (2009). The relation of α-linolenic acid to the risk of prostate cancer: A systematic review and meta-analysis. *The American Journal of Clinical Nutrition*, 89(5), 1558S–1564S.

Surette, M. E. (2008). The science behind dietary omega-3 fatty acids. *CMAJ: Canadian Medical Association Journal = Journal de l'Association Medicale Canadienne*, 178(2), 177–180.

Swanson, D., Block, R., & Mousa, S. A. (2012). Omega-3 fatty acids EPA and DHA: Health benefits throughout life. *Advances in Nutrition*, 3(1), 1–7.

Vannice, G., & Rasmussen, H. (2014). Position of the academy of nutrition and dietetics: Dietary fatty acids for healthy adults. *Journal of the Academy of Nutrition and Dietetics*, 114(1), 136–153.

Virsangbhai, C. K., Goyal, A., Tanwar, B., & Sihag, M. K. (2019). Potential health benefits of conjugated linoleic acid: An important functional dairy ingredient. *European Journal of Nutrition & Food Safety*, 11(4), 200–213.

4

Properties
Physical, Chemical and Functional

Vandana Chaudhary and Priyanka Kajla

Contents

4.1 INTRODUCTION

The term "fatty acids" refers to long carbon chains with a carboxyl group at one end and a methyl group at the other end, particularly in fats and oils and also as a chief component of cell membranes. These are significant components of the human body, having functional, structural and biological roles. Fatty acids, either free or as constituents of complex lipids, play various crucial roles in metabolism:

- Storage of energy
- Constituents of membranes
- Cellular fuel sources
- Major component of lipids as well as hormones
- Gene expression
- Regulate intracellular signal pathway (Calder PC, 2015)
- Function as precursor of cytokine
- Act as sources of polyunsaturated fatty acids (PUFA)
- PUFA plays an important role in cardiovascular and neuro protection
- PUFA helps in diminishing the oxidative stress
- Salts of long-chain fatty acids (palmitic and stearic) have cleansing action and are known as soaps and detergent
- Used as biomarkers to identify certain pathological conditions like polycystic ovarian disorder (PCOD)

4.2 CLASSIFICATION OF FATTY ACIDS

Fatty acids are the basic units of oils and fats. They have a parent chain of carbon atoms. The fatty acids may vary according to the units of carbon atoms and the kind of bonding between two subsequent carbon atoms. These are classified in multiple manners. They may be segregated on the account of the presence and number or quantity of double bonds as saturated and unsaturated fatty acids. Unsaturated fatty acids have at the minimum one double bond incorporated in their hydrocarbon chain.

A fatty acid is termed monounsaturated if it embodies a sole double bond, and it is designated as a polyunsaturated fatty acid (PUFA) if it includes more than one double bond (Figure 4.1).

A number of physiological properties of unsaturated fatty acids are chiefly determined by the location of first unsaturation in relation to the end position and not the carboxyl group. The carbon adjacent to the

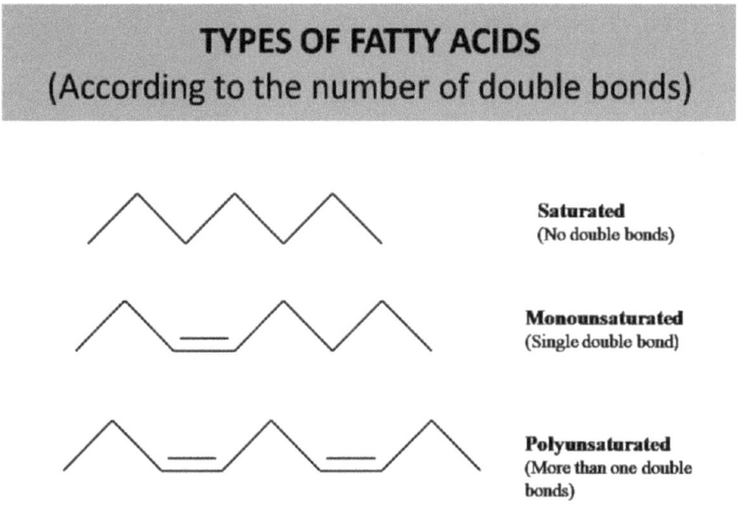

Figure 4.1 Types of fatty acids according to number of double bonds

carboxyl group is titled as an "α" carbon, the subsequent carbon atom as "β" and so on. The rearmost carbon is tagged as a "ω."

PUFAs can be further subdivided on the basis of the site of the first double bond with reference to the methyl terminal of the fatty acid chain as omega (ω)-3 and ω-6 fatty acids. Such fatty acids are also christened as essential fatty acids (EFAs), as humans and other animals cannot synthesize them. These have to be ingested in other forms for good health. In the human body, these EFAs are utilized to produce substances which help in the regulation of blood pressure and neuronal firing, the production of hormones and so on. These are also metabolized into other compounds like lipoxins and eicosanoids, which impart an anti-inflammatory effect.

4.3 PROPERTIES OF FATTY ACIDS

4.3.1 Physical Properties

The fundamental structure of fatty acids consists of a hydrophobic (oil-loving) hydrocarbon chain at one end and a hydrophilic polar group at the other end (Figure 4.2), thus endowing fatty acids and their derivatives with peculiar properties.

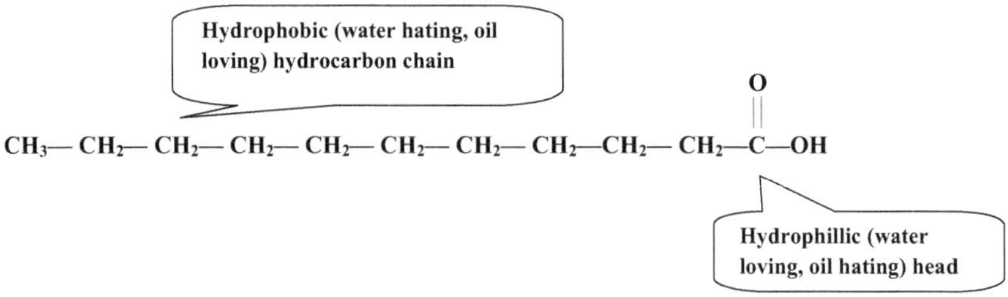

Figure 4.2 Structure of fatty acid

4.3.1.1 Melting Point

Melting point signifies the degree or intensity of heat whereupon the solid form transforms to the liquid state. The structure of a fatty acid greatly affects its melting point: for example, fatty acids with cis double bonds and branched chains have lower melting points than their equivalent saturated fatty acids. Besides this, the melting point is also dependent on the number of carbon atoms in the fatty acid parent chain. Chains with an odd number of carbon atoms have higher melting points in comparison to chains with an even number of carbon atoms. Another factor influencing the melting point of fatty acids is the quantity of double bonds contained in the chain. The more double bonds, the lower will be the melting point. The underlying cause for this kind of behavior can be attributed to molecular geometries. The existence of double bonds in the fatty acid leads to bends in the molecule. As a result of this, the molecules cannot stack well, leading to weaker intermolecular interaction as compared to saturated ones. Consequently, the melting point declines from stearic acid (70°C) to oleic acid (13°C) to linoleic acid (–5°C) to linolenic acid (–11°C) and finally to arachidonic acid (–50°C) (Virtual Chembook). In other words, the more double bonds, the more liquid it will be. Saturated fats in general will be solid at room temperature, whereas monounsaturated oils remain liquid to lower temperatures, and polyunsaturated fatty acids stay liquid beneath the freezing point of water.

4.3.1.2 Oxidation Stability

Oxidation stability is one of the crucial quality criteria for determining the usefulness of edible oils during processing and shelf life determination. Fatty acids having single bonds are very stable, whereas unsaturated acids are more prone to oxidation. The greater the number of double bonds and the proximity to heat, light, pigments, antioxidants and pro-oxidants like metal ions, especially iron, the more vulnerable the fatty acid will be to oxidation. The carbon atom adjoining the double bond is more reactive, as the hydrogen attached to it can be abstracted by the reactive oxygen radicals, which will start a chain reaction. This results not only in loss of bioactivity but also in the production of objectionable off-odors and off-flavors. The rate of oxidation of linolenic acid is highest, followed by linoleic and oleic acids (Maszewska et al., 2018). The effects of the degree of unsaturation on the stability of oil on comparing stored regular and low-linolenic canola oils were 12.5% and 2.5% of 18:3, respectively (Malcolmson, Vaisey-Genser, Przybylski, Ryland, Eskin and Armstrong, 1996).

To illustrate, the pace of oxidation of oleate (*cis* 18:1), linoleate (*cis* 18:2) and linolenate (*cis* 18:3) have been claimed to be 1:40–50:100 owing to the amount of oxygen taken and 1:12:25 due to the formation of hydroperoxides (Frankel, 2005). The ω-3 FAs DHA and EPA are even more sensitive to oxidation, signifying that they are incredibly challenging to be incorporated within foods. Berlitz et al., 2009 reported that the relative oxidation rate of PUFA constituting α-linolenic acid (an ω-3 FA) is virtually 2500 times more than saturated stearic acid. Hence, unsaturated fatty acids such as EPA, DHA, AL and CLA should be managed under an inert gas atmosphere and should be safeguarded against oxidants and compounds initiating the formation of free radicals.

4.3.1.3 Smoke Point

The smoke point is the temperature at which oil or fat under specific conditions starts producing visible bluish smoke. It helps in the characterization of oils regarding suitability for frying. Smoke point values can fluctuate relying on various factors including oil quality, the free fatty acid (FFA) content of the oil, the amount of oil used, light, oxygen and source of light. The higher the FFA content, the lower the smoking point will be. This shows that unrefined oils which have not undergone refining have high concentrations of free acids and they may be rendered unsuitable for frying.

The optimal temperature for frying is ~180°C. Hence, a cooking oil should have a high smoke point and an optimum balance of ω-3 and ω-6 fatty acids (Vieira et al., 2015). For cooking to be carried out at high temperatures, cooking oils having high smoke points must be selected (Table 4.1). Oils with high concentration of ω-3 fatty acids are preferred for salad dressings or low-temperature cooking, since they boost healthy cells and augment cardiovascular health.

Table 4.1 Smoke Point of Certain Oils Rich in Polyunsaturated Fatty Acids

Cooking Oil/Fat	Smoke Point (°C)
Safflower oil (unrefined)	107
Flaxseed oil (unrefined)	107
Walnut oil (unrefined)	160
Sesame oil	210
Cottonseed oil	216
Palm oil	232
Rice bran oil	254
Avocado oil	271
Sunflower	227
Soya bean oil	232

4.3.1.4 Refractive Index (RI)

The refractive index evaluates the light-bending capacity as a ray of light moves from one medium into another. The RI of fats and oils is a basic value that relates to the number of double bonds, molecular weight, composition, chain length and degree of conjugation. It is directly related to the number of carbon atoms and the length of the fatty acid chain and inversely proportional to temperature. Consequently, the refractive index ascertains the probability in reference to the unsaturation and chain length of the sample. It can also increase with exposure to light and heat. In general, the refractive index value varies between 1.447 and 1.482 (Arya et al., 1969). Mostafa et al., 2013 reported that the refractive indexes of flaxseed oil, canola oil and olive oil were 1.481, 1.467 and 1.462, respectively, at 25°C.

4.3.1.5 Solubility

In regard to solubility, the fatty acids in their un-dissociated or acidic form are generally insoluble in water. In contrast, their tendency to solubilize in water increases with the generation of salts in combination with sodium or potassium. This is attributed to the existence of carboxylic acid at the end of PUFA, which reacts with bases such as alkali metal hydroxides and amines to form ionic salts having discernible ionic behavior, making them soluble in water. The presence of heavy metals like lead and mercury leads to the formation of salts having prominent covalent characteristics, thus limiting their water solubility. Basically, the water solubility of fatty acids, in particular long-chain FA, is perpetually very complex to judge, as it is notably affected by hydrogen ion concentration and the tendency of fatty acids to consociate with each other, which leads to the fabrication of spherical clusters known as *micelles*.

4.3.1.6 Specific Heat

The amount of energy/heat absorbed by a substance to increase its temperature by 1 degree is termed its specific heat. Fats in the liquid state have approximately double the specific heat values compared with solid fats. In a study conducted by Santos et al. (2005), it was deduced that specific heat has a tendency to increase with an increase in the saturation of fatty acids. Another study, by Sathivel (2005), pointed that the specific heat capacities for salmon and pollock oils ranged from 0.8 to 2.3 and from 0.6 to 3.5 (j/g/°C), respectively. Contreras-Gallegos et al. (2017) showed that extra-virgin olive oil and avocado oil, constituting mainly of monounsaturated acids, showed higher specific heat values, and in contrast extra-virgin flaxseed oil, consisting mainly of polyunsaturated acids, exhibited reduced specific heat values.

4.3.1.7 Viscosity

Viscosity assumes a very noteworthy part in determining the rheological properties of oils. It signifies the internal opposition of a fluent to get deformed under shear stress. Generally it is apprehended as defiance to pouring or thickness. The viscosity of oils is measured as well in two ways: absolute viscosity and kinematic viscosity (Diamante and Lan, 2014). Absolute viscosity is the resistance of oil to shear and flow as a result of internal friction, whereas kinematic viscosity may be defined as the resistance to shear and flow due to gravity. It has been proven that absolute viscosity is inversely proportional to temperature: as temperature increases, viscosity declines. Absolute viscosity is primarily considered while designing heat exchangers, pumping requirements and so on. With increases in fatty acid size, viscosity increases; meanwhile, viscosity decreases with increased unsaturation. Research conducted by Yalcin et al., 2012 to study the implications of the type of oil and constituent fatty acids on the deformation and flow of vegetable oils revealed that there exists an interrelation between viscosity and the concentration of monounsaturated and polyunsaturated fatty acids in oils. A decline in oil viscosity was witnessed with a surge in the polyunsaturated fatty acid proportion and a decrease in the monounsaturated fatty acid concentration. These perceptions affirm the findings determined by Fasina et al., 2006 that the proportions of MUFA and PUFA were highly correlated ($R^2 > 0.82$) with the viscosities of the oil samples.

4.3.1.8 Specific Gravity

Specific gravity signifies the heaviness of fats and oil compared to that of water. Generally, the specific gravity of oils increases with unsaturation levels. Recently, Katkade et al., 2018 revealed that the specific gravity of safflower oil was 0.917. The higher specific gravity value of this oil was attributed to the different proportion of C18 PUFA. It was also reported that temperature has a robust impact on specific gravity, with a downtrend when the temperature is escalated.

4.3.2 Chemical and Functional Properties

4.3.2.1 Hydrogenation

Hydrogenation is the process by which unsaturated fatty acids in vegetable oils, as well as marine or animal fats, are converted to saturated fatty acids by the addition of hydrogen at the points of unsaturation. This process is carried out at high pressure and/or temperature and in the presence of catalysts. Hydrogenation has long been carried out to improve the oxidative stability of oils and to modify the melting characteristics as well as the solid content of oils. In addition, it is an exothermic reaction and helps in increasing the melting point and decreasing the iodine value. Polyunsaturated fatty acids, e.g., linolenic acid (C18:3), undergo hydrogenation much more promptly in comparison to oleic acid (C18:1) and linoleic acid (C18:2) than what is seen for linoleic to oleic acid or oleic acid to stearic acid (C18:0).

$$\text{Linolenic} \xrightarrow{(+)2H} \text{Linoleic} \xrightarrow{(+)2H} \text{Oleic} \xrightarrow{(+)2H} \text{Stearic}$$

During the hydrogenation process, hydrogen molecules after being engrossed onto the nickel are segregated into two individual atoms. Fatty acids are also imbibed on the nickel with the help of their double bond/s; thereafter the hydrogen atoms originated in the initial steps are attached to the double bond leading to the formation of a semi-hydrogenated intermediate. On further addition of a second hydrogen atom, the double bond becomes saturated. Essential fatty acids are hydrogenated to a smaller extent, but partial hydrogenation of linoleic and linolenic acid imparts oxidative stability. Therefore, the oils enriched with these essential fatty acids can be utilized as salad oil and also as shortening for baked products.

4.3.2.2 Halogenation

PUFA, when treated with halogens like bromine and chlorine under mild conditions, undergoes an addition reaction consuming two halogen atoms per double bond, yielding a halogenated derivative of the respective fatty acid. The number of halogen atoms added to a specific fatty acid is an indicator of the degree of unsaturation.

$$CH_3CH_2CH{=}CHCH_2CH{=}CH\left(CH_2\right)_7 COOH + 3Br_2$$

$$\rightarrow CH_3CH_2CHBr\text{-}CHBr\ CH_2\ CHBr\text{-}CHBr\left(CH_2\right)_7 COOH$$

Venturi and Venturi (2014) found that iodide may be used as an endogenous antioxidant to reduce the passive oxidative damage to DHA confined in the brain and retina, thus enhancing the functioning of these crucial organs of the human body. It also found that iodine is more potent in preventing oxidative damage than vitamin C (ascorbic acid), glutathione and vitamin E (tocopherols), which may be attributed to its small size or because of its solubility in lipids. Sturza (2006) suggested that elemental iodine could be added to sunflower oil, which may be further used as either a food additive or an injectable dietary supplement. In addition, brominated vegetable oils are commonly used as an emulsifying agent in soft drinks, especially in citrus-flavored drinks.

4.3.2.3 Iodine Value (IV)

Fatty acids containing double bonds can be transformed into saturated fatty acids by undergoing addition reactions. Relying on the extent of unsaturation, the fatty acids have the ability to combine with halogens or oxygen to form saturated fatty acids. The level of unsaturation in fatty acids can be checked by employing various methods, and iodine value (IV) is one of them. IV may be characterized as grams of iodine absorbed by 100 g of fat. Larger iodine value denotes elevated unsaturation. As oxidative stability and melting point are functions of the degree of unsaturation, IV assists in the assessment of these quality factors. The higher the iodine value, the greater the unsaturation and the susceptibility towards oxidation. Fish oils are considered to be highly beneficial from a nutritional point of view, as they have higher levels of unsaturated fatty acids, in particular EPA and DHA, and they exhibit higher iodine values (Ogundiran and Ojo, 2012). The iodine values of various oils having abundant amounts of EFA, namely, linseed, cod liver, tung and walnut oil were reported as 136–178, 145–180, 163–173 and 12–155, respectively (Thomas, 2002). Another study by Imran et al. (2016) documented that chia seed oil rich in polyunsaturated fatty acids recorded iodine values of 182.83 for autoclaved samples and 193.42 for raw chia seed samples.

4.3.2.4 Rancidity

Fats and oils over a long period of time and/or during exposure to moisture, light, air and heat endure oxidation and hydrolysis reactions, thus developing an unpleasant odor and becoming rancid. There are two basic types of rancidity: oxidative, which takes place in the presence of oxygen or UV rays, and hydrolytic/enzymatic, occurring in the presence of moisture. Hydrolytic rancidity is predominantly consummated through naturally occurring enzymes in animal fat (lipase) and plant oils such as cyclooxygenase and lipoxygenase. Another quality degradation process may be microbial rancidity. Microorganisms like bacteria, yeast and molds may use their enzymes to break down compounds present in oils, making them more vulnerable to developing objectionable flavors, thus affecting consumer acceptability.

The paramount stimulus of rancidity is the oxidation of unsaturated fatty acids by the free radical mechanism (Figure 4.3). Usually it occurs in

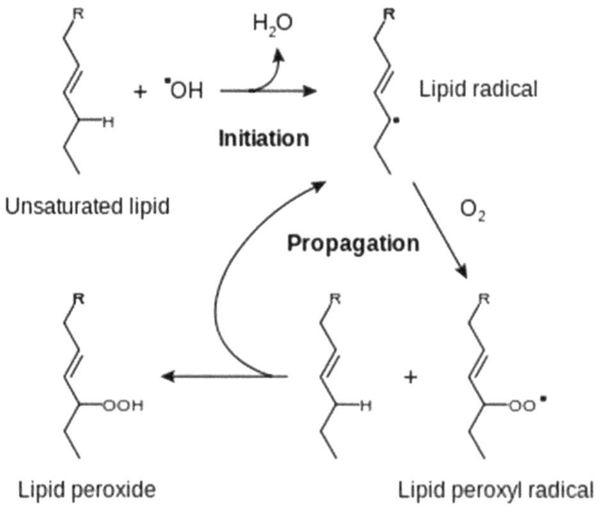

Figure 4.3 Fatty acid oxidation mechanism (Quora, 2017. https://www.quora.com/Why-are-unsaturated-fatty-acids-less-stable)

three phases. In the initial phase, the reaction is triggered by the combination of a free oxygen molecule with the unsaturated fatty acid, producing free radical and hydroperoxides. The products of this reaction are highly unstable and thus reactive in nature. The second phase is described as *propagation*. Unstable and highly reactive free radical species formed in the first phase initiate reaction with other fatty acids, consequently contributing to a chain reaction, also known as auto-oxidation. The last stage of oxidation of fatty acids is *termination*, which is marked by dragging or pausing of reaction and the formation of stable compounds, for instance aldehydes, alcohols or ketones.

Fatty acids on being oxidized yield two types of by-products: primary (hydroperoxides and peroxides) and secondary (alcohols, carboxylic acids, aldehydes). The peroxide value test is usually employed for measuring primary oxidation, and the anisidine value test is used for secondary oxidation.

Oils with greater concentrations of fatty acids having one or more double bonds are highly liable to turn rancid. This is because less energy is required to break the double bonds, thus rendering them unstable, which allows oxygen molecules to react with fatty acids at the point of unsaturation. Moreover, oxygen is eight times as soluble in oils than in water. One of the chief causes of rancidity is the release of the volatile, noxious-smelling compound malonaldehyde due to the oxidative cleavage double bonds of unsaturated fatty acid.

As a result of peroxidation of ω-6 and ω-3 PUFAs, hydroxy alkenals such as 4-hydroxy-2-nonenal (HNE) and 4-hydroxy-2-hexenal (HHE), respectively, are generated as by-products, which in turn produces undesirable flavors resulting in lowering the nutritional value of PUFA-rich foods and also decreasing shelf life (Tao, 2015).

4.3.2.5 Peroxide Value

Peroxide value (PV) is a very crucial assay for determining the state of oxidation in oils and fats. Fatty acids having more than one double bond are extremely sensitive to being oxidized, leading to the production of hydroperoxides and thus enabling the estimation of oxidation status specifically in the primary state of oxidation, also denoted by the term "induction stage." The peroxide value is routinely expressed as the milliequivalents of peroxides per kilogram of fat but while expressing in terms of SI units it is halved and conveyed as millimoles of active oxygen per kilogram of fat.

GOED (Global Organization for EPA and DHA Omega-3s) is an organization comprised of generators of omega-3 long-chain polyunsaturated fatty acids including products, along with relevant industries. GOED has devised standards for its allies that produce omega-3 long-chain polyunsaturated fatty acid–fortified oil. The oil produced should strictly adhere to the following requirements: peroxide value should be less than 5 milliequivalents of oxygen per kilogram, the anisidine value should be less than 20, and the combination of both (denoted "TOTOX") should be less than 26 (GOED, 2015). A study published in New Zealand proclaimed that 69% of the 32 products containing fish oil in the market (Albert et al., 2015) contained lower EPA/DHA ratio compared to the values indicated on the label; in addition, most of the products also surpassed the maximal oxidation cap conveyed by GOED. Going by the results obtained, a relationship was found between the stage of oxidation and the low ratio of EPA/DHA, and it was implied that oxidation may be termed as the culprit that could add to the depression in the EPA/DHA ratio. It was also communicated that in New Zealand the majority of health supplements incorporating fish oil were not in accordance with the set standards and moreover did not meet the mentioned content of omega-3 fatty acid on the labels and were exceedingly oxidized (Bannenberg, 2017).

4.3.2.6 Anisidine Value

The oxidation of PUFA engenders the release of a number of breakdown products, firstly primary oxidation by-products like free fatty acids and dienes; secondary products such as aldehydes (primarily 2-alkenals and 2,4-alkadienals), carbonyls and ketones; and tertiary products. The anisidine value (AV) parameter measures the secondary oxidation by-products, originated as a result of decomposition of hydroperoxides. As AV provides a quantifiable indication of secondary oxidation products, it can be used independently or in combination with PV for judging the freshness of oils and fats (Yang and Boyle, 2016). GOED advises a maximum limit for AV of 20 milliequivalents per kilogram. As the oxidation of oils containing essential fatty acids proceeds, the EPA and DHA percentages further attenuate, thereby reducing their effectiveness (Martinsen, 2019).

4.3.2.7 Stereomutation

Stereomutation may be referred to as transformation of a molecule configuration by either physical or chemical methods at a stereogenic unit. Predominantly, fatty acids occur in their cis geometry but can be changed intentionally or accidentally into trans fatty acids (Dhaka et al., 2011). In general, the compound is an equilibrium concoction of all probable isomers; however, trans isomers, being more stable thermodynamically, exist in the ratio 4:1. For instance, linoleic acid, a PUFA, can generate a combo of four probable isomers namely cis, cis acid; trans, trans isomer; and two cis trans isomers. When PUFAs are exposed to elevated temperatures like that in the process of deodorization, they undergo stereomutation. It has been estimated that linolenic acid is isomerized about 14 times faster than linoleic acid (Gunstone, 2004). Experimental results achieved by Ferrari et al. (2012) indicated that when fish oils undergo the deodorization process, high temperatures convert EPA (PUFA) into trans-isomers.

4.4 CONCLUSIONS

Essential fatty acids, in particular omega-3 fatty acids, are largely utilized by the pharmaceutical industry due to their multitudinous health benefits. These fatty acids, due to the presence of numerous double bonds in chemical structures, are very vulnerable to degradation during handling, processing and storage. Therefore, proper understanding of the physical and chemical properties of essential fatty acids is a must to improve their functionality in processed food products.

Because of their higher susceptibility to temperature and oxygen, these fatty acids exhibit poor quality attributes in food products. Hence, it is required to conduct more research to encourage the wider utilization/incorporation of essential fatty acids in processed foods to increase their dietary intake. It is recommended that knowledge of functional properties might serve as a tool for tailoring essential fatty acids as per processing requirements.

REFERENCES

Albert B.B., Derraik J.G., Cameron-Smith D., Hofman P.L., Tumanov S., Villas-Boas S.G., Garg M.L. 2015. Cutfield WS science report Jan 21, vol. 5, p. 7928.

Arya S.S., Ramanujam S., Vijayaraghavan P.K. 1969. Refractive index as an objective method for evaluation of rancidity in edible oils and fats. *Journal of the American Oil Chemists' Society* 46(1): 28–30.

Bannenberg G., Mallon C., Edwards H., Yeadon D., Yan K., Johnson H., Ismail A. 2017. Omega-3 long-chain polyunsaturated fatty acid content and oxidation state of fish oil supplements in New Zealand. *Scientific Reports* 7(1): 1488.

Berlitz H.D., Grosche W., Schieberle P. 2009. Lipids. In: *Food Chemistry*, eds. Berlitz H.D., Grosch W., Schieberle P. Berlin, Germany: Springer, pp. 158–247.

Calder P.C. 2015. Functional roles of fatty acids and their effects on human health. *Journal of Parenteral and Enteral Nutrition* 39(1): 18S–32S.

Contreras-Gallegos E., Domínguez-Pacheco F.A., Hernández-Aguilar, C. et al. 2017. Specific heat of vegetable oils as a function of temperature obtained by adiabatic scanning calorimetry. *Journal of Thermal Analysis and Calorimetry* 128: 523–531.

Dhaka V., Gulia N., Ahlawat K.S., Khatkar B.S. 2011. Trans fats: Sources, health risks and alternative approach - A review. *Journal of Food Science and Technology* 48(5): 534–541.

Diamante L.M., Lan T. 2014. Absolute viscosities of vegetable oils at different temperatures and shear rate range of 64.5 to 4835 s^{-1}. *Journal of Food Processing.* Article ID 234583, 6 pages. doi:10.1155/2014/234583.

Fasina O.O, Hallman H, Craig-Schmidt M, Clements C. 2006. Predicting temperature-dependence viscosity of vegetable oils from fatty acid composition. *Journal of the American Oil Chemists' Society* 83(10): 899–903.

Ferreri C., Grabovskiy S.A., Aoun M., Melchiorre M., Kabal'nova N., Feillet-Coudray C., Fouret G., Coudray C., Chatgilialoglu C. 2012. Trans fatty acids: Chemical synthesis of eicosapentaenoic acid isomers and detection in rats fed a deodorized fish oil diet. *Chemical Research in Toxicology* 25(3): 687–694.

Frankel E.N. 2005. *Lipid Oxidation.* 2nd ed. Bridgewater (UK): The Oily Press.

Global Organization for EPA and DHA. 2015. Omega-3. GOED Voluntary Monograph - v.5. http://www.goedomega3.com/index.php/files/download/350.

Gunstone F.D. 2004. Chemical properties related to unsaturated centres in the chemistry of oils and fats: Sources, composition, properties and uses, pp. 146–188.

Imran M., Nadeem M., Manzoor M.F., Javed A., Ali Z., Akhtar M.N., Ali M., Hussain Y. 2016. Fatty acids characterization, oxidative perspectives and consumer acceptability of oil extracted from pre-treated chia (*Salvia hispanica* L.) seeds. *Lipids in Health and Disease* 15(1): 162.

Katkade M.B., Syed H.M., Andhale R.R., Sontakke M.D. 2018. Fatty acid profile and quality assessment of safflower (*Carthamus tinctorius*) oil. *Journal of Pharmacognosy and Phytochemistry* 7(2): 3581–3585.

Malcolmson, L.J., Vaisev-Genser, M., Przybylski, R. et al. 1996. Characterization of stored regular and low-linolenic canola oils at different levels of consumer acceptance. *Journal of the American Oil Chemists' Society* 73: 1153–1160 (1996)

Martinsen B. 2019. Making sense of fish oil oxidation values. Omega 3 Innovations. https://omega3innovations.com/blog/making-sense-of-fish-oil-oxidation-values/.

Maszewska M., Florowska A., Dłużewska Elżbieta, Wroniak Małgorzata, Marciniak-Lukasiak Katarzyna, Żbikowska Anna 2018. Oxidative stability of selected edible oils. *Molecules* 23(7): 1746.

Mostafa R.A., Moharram Y.G., Attia R.S., El–Sharnouby R.S. 2013. Formulation and characterization of vegetable oil blends rich in omega-3 fatty acids. *Emirates Journal of Food and Agriculture* 25(6): 426–433.

Ogundiran Mary B., Ojo Adeniyi S. 2012. Determination of fat contents, iodine values, Trace and toxic metals in commonly consumed frozen fish in Nigeria. *American Journal of Food Technology* 7(1): 34–42.

Ophardt C.E. 2003. Virtual Chembook [online]. Available at: http://www.elmhurst.edu/~chm/vchembook/551fattyacids.html.
Quora. 2017. Why are unsaturated fatty acids less stable? https://www.quora.com/Why-are-unsaturated-fatty-acids-less-stable.
Santos J.C.O., Conceição M.M., Santos M.G.O., Dantas J.P., Conceio M.M., Athaide-Filho P.F., Souza A.G. 2005. Comparative study of specific heat capacities of some vegetable oils obtained by DSC and microwave oven. *Journal of Thermal Analysis and Calorimetry* 79(2): 283–287.
Sathivel S. 2005. Thermal and flow properties of oils from salmon heads. *Journal of the American Oil Chemists' Society* 82(2): 147–152
Sturza R. 2006. Iodination of vegetable oils as a method for correcting iodine deficiency. *Chemistry Journal of Moldova* 1(1): 65–72.
Tao L. 2015. Oxidation of polyunsaturated fatty acids and its impact on food quality and human health. *Advances in Food Technology and Nutritional Sciences: Open Journal* 1(6): 135–142.
Venturi S., Venturi M. 2014. Iodine, PUFAs and iodolipids in health and diseases: An evolutionary perspective. *Human Evolution* 29(1–3): 185–205.
Vieira S.A., McClements D.J., Decker E.A. 2015. Challenges of utilizing healthy fats in foods. *Advances in Nutrition* 6(3): 309S–317S.
Yalcin H., Toker O.S., Dogan M. 2012. Effect of oil type and fatty acid composition on dynamic and steady shear rheology of vegetable oils. *Journal of Oleo Science* 61(4): 181–187.
Yang X., Boyle R.A. 2016. Sensory evaluation of oils/fats and oil/fat–based foods. In: *Oxidative Stability and Shelf Life of Foods Containing Oils and Fats*, Eds. Hu M., Jacobsen C., Cambridge, MA: Academic Press and AOCS Press, pp. 157–185.

5

Mechanism of Action of Essential Fatty Acids

Sneh Punia, Sanju Bala Dhull, Anil Kumar Siroha,
Kawaljit Singh Sandhu and Vandana Chaudhary

Contents

5.1 INTRODUCTION

For the survival and proper functioning of our body organs, our daily food consumption plays an important role by providing nutrients. Along with supplying energy, foods have the potential to fight against degenerative and chronic diseases. The presence of bioactive compounds in plant sources is the key factor for the prevention of diseases and individual health. Phenolic acids, flavonoids, carotenoids and n-3, n-6 polyunsaturated fatty acids (PUFAs) are the primary bioactive compounds. Polyunsaturated fatty acids play an important part in the structure and function of cellular membranes and are precursors of lipid mediators (Carlier et al., 1991). Among all polyunsaturated fatty acids, only linoleic acid (LA) and α-linolenic acid (ALA) are characterized as essential fatty acids (EFAs). Omega-6 (n-6) fatty acids are depicted by LA and omega-3 (n-3) fatty acids by ALA (Simopoulos, 2016). LA and ALA need to be provided in the diet, as they are required for human biological activities but cannot be synthesized by the body because of the lack of Δ15-desaturase, which is required for synthesizing FAs. ALA acid is the dietary-essential precursor of EPA and DHA; the main site for the synthesis of EPA and DHA from ALA is the liver, and the rate of conversion is about 8% for EPA and 1% for DHA (Laye et al., 2018). ALA, EPA and DHA are the primary omega-3 fatty acids, and they have their double bond in the ω-3 (carbon-3) position and are in cis configuration. Both EPA and DHA play key roles in preventing and treating poor health and well-being and have beneficial effects in the prevention of cardiovascular disease (CVD) (Balk and Lichtenstein, 2017), inflammation (Laye et al., 2018), hypertension (Colussi et al., 2016) and cancer (D'Eliseo and Velotti, 2016). Omega-3 and omega-6 fatty acids are very crucial for the synthesis of eicosanoids such as prostaglandins (PGs), leukotrienes (LTs), thromboxane (TX), prostacyclin (PGI) and lipoxins, which play an essential role in vascular physiology. As reported by many studies, associated health benefits of essential fatty acids are dependent on metabolism, absorption and bioavailability in the human body; therefore this chapter focuses on all these parameters briefly.

5.2 METABOLISM AND ABSORPTION

In the human diet, about 95% of fat is in the form of triacylglycerols (TAG). To produce 2 free fatty acids (FFAs) and 2-monoacylglycerol (MAG), TAGs require lipolysis of sn-1 and sn-3 positions for efficient solubilization and uptake into the enterocyte (Carlier et al., 1991). Hydrolysis of dietary lipids in humans is catalyzed by lipases from gastric and pancreatic origins (Christensen et al., 1995). This hydrolysis takes place at the low pH of the stomach lumen in the absence of bile salts. This emulsification of lipids enhances further hydrolysis by pancreatic lipid hydrolases in the small intestine, since these enzyme activities depend on the emulsification of dietary fat (Carlier et al., 1991). During digestion of fat, free fatty acids and monoacylglycerols are released and absorbed in small intestine. Further digestion of emulsified fat takes place in the proximal part of the small intestine (end of the duodenum and the jejunum). In the intestinal mucosa cells, FFA is re-esterified to triacylglycerols, which are transported to the circulation as part of chylomicrons. Chylomicrons are lipoprotein molecules which transport dietary lipids from the intestines to other locations in the body. In the circulation, fatty acids are transported bound to albumin or as part of lipoproteins (Rustan and Drevon, 2001). FFA is taken up into cells mainly by protein transporters in the plasma membrane and is transported intracellularly (Figure 5.1). FFAs are then activated (acyl-CoA) before they are shuttled via acyl-CoA-binding protein (ACBP) to mitochondria or peroxisomes for β-oxidation (and formation of energy as ATP and heat) or to the endoplasmic reticulum for esterification to different classes of lipid. Acyl-CoA or certain FFAs may bind to transcription factors that regulate gene expression or may be converted to signal molecules (eicosanoids). Glucose may be transformed to fatty acids (lipogenesis) if there is a surplus of glucose/energy in the cells (Figure 5.2).

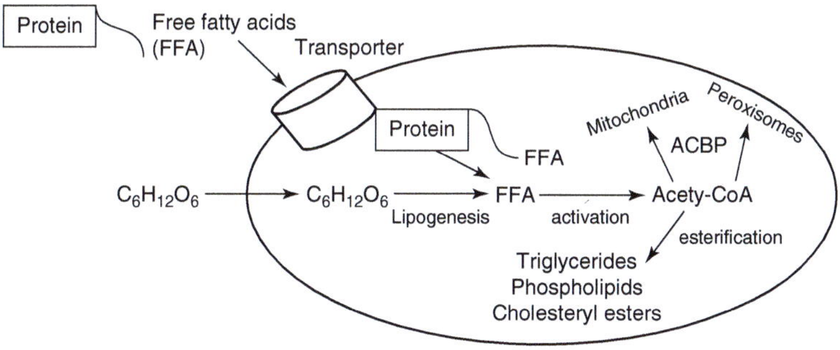

Figure 5.1 Metabolism of fatty acids (FFA, free fatty acids; ACBP, acyl-CoA binding protein)

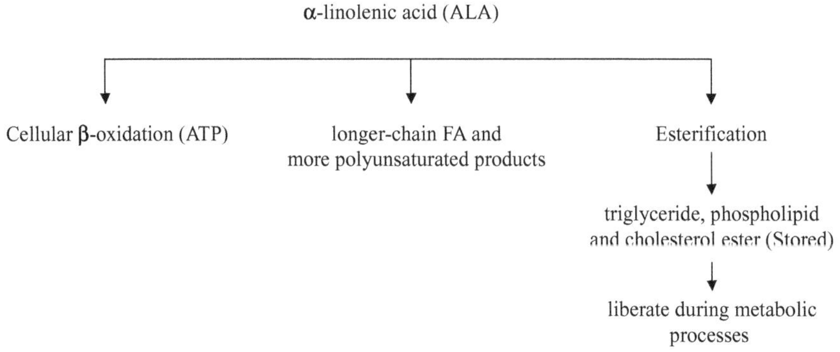

Figure 5.2 Metabolism of ALA fatty acids (overview)

5.2.1 Essential Fatty Acids

ALAs are consumed as triglycerides, digested and absorbed in the small intestine and then transported to body tissues through plasma (Kaur et al., 2014). ALA is the simplest member of biologically active fatty acids and may be converted to EPA and DHA, which are more biologically active (Figure 5.3). For efficient digestion and absorption of fatty acids, firstly fat is emulsified by bile secretion into the small intestine. ALA undergoes (1) esterification (for triglycerides and phospholipids, important functioning of n-6 and n-3 fatty acids), (2) oxidation and (3) formation of longer-chain and saturated products. Mammals are not able to synthesize longer-chain ALAs, but obtain shorter-chain ALAs (18 carbons and 3 double bonds) to form EPA (20 carbons and 5 double bonds) and then DHA (22 carbons and 6 double bonds). Stearidonic acid is the first metabolite synthesized from ALA by Δ6-desaturase, and then stearidonic acid may be elongated to eicosatetraenoic acid. Further desaturation by Δ5-desaturase yields EPA and DPA. During conversion of ALA into DHA, a number of n-3 fatty acids are formed through desaturation and elongation reactions (Figure 5.2). DPA desaturation at the Δ4 position to produce DHA directly or as reported by Sprecher (2000), which involves the elongation and desaturation of DPA into tetracosapentaenoic acid (by elongase) and tetracosahexenoic acid (by desaturase) and finally to DHA (by peroxisomal oxidation) and α-linolenic acid (ALA).

Linoleic acid may be metabolized to long-chain members of the n-6 family by the insertion of additional double bonds during consecutive elongation and desaturation mechanisms (Patterson et al., 2012) (Figure 5.4). The

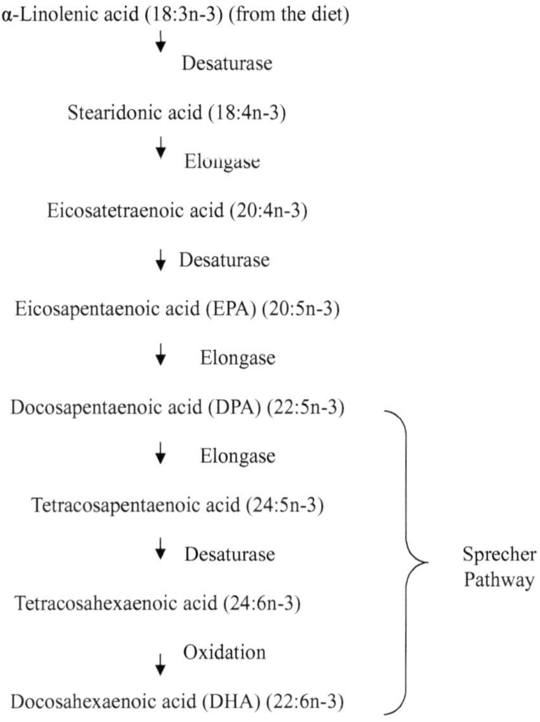

Figure 5.3 Metabolism of ALA fatty acids into EPA and DHA

initial rate-limiting desaturation of LA to GLA is catalyzed by the enzyme Δ6-desaturase (Stoffel et al., 2008). Elongation then takes place to convert GLA to DGLA, by elongation of very-long-chain fatty acids (ELOVL) 5, and finally a cycle of elongation and desaturation by Δ5-desaturase (FADS1) generates arachidonic acid (AA) (Moon et al., 2009).

Long-chain PUFAs, EPA and DHA are further metabolized by the enzymes cyclooxygenase and lipoxygenase to eicosanoids (Figure 5.5), prostaglandins and leukotrienes. Eicosanoids are important for several cellular functions such as platelet aggregability (ability to clump and fuse), chemotaxis (movement of blood cells) and cell growth. Eicosanoids are rapidly produced and degraded in the cells where they execute their effects (Rustan and Drevon, 2001). Among these eicosanoids, E2 series prostaglandins, B4 leukotrienes derived from linoleic acid are the key metabolites responsible for many inflammatory diseases like cardiovascular diseases and arthritis, while eicosanoids and E3 series prostaglandins

Linolic acid (18:2n-6)

↓ Desaturase

Gamma-linolenic acid (GLA,18:3n-6)

↓ Elongase

Dihomo-gamma-linolenic acid (DGLA, 20:3n-6)

↓ Desaturase

Arachidonic acid (AA, 20:4n-6)

↓ Elongase

Docosatetraenoic acid (DTA, 22:4n-6)

↓ Elongase

Tetracosatetraenoic acid (TTA, 24:4n-6)

↓ Desaturase

Tetracosapentaenoic acid (TPA, 24:5n-6)

↓ Oxidation

Docosapentaenoic acid (DPA, 22:5n-6)

Figure 5.4 Metabolism of LA fatty acids

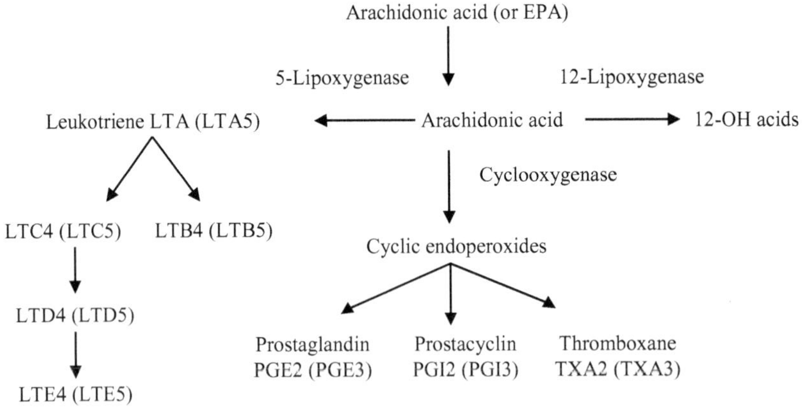

Figure 5.5 Synthesis of eicosanoids from arachidonic acid or eicosapentaenoic acid (EPA)

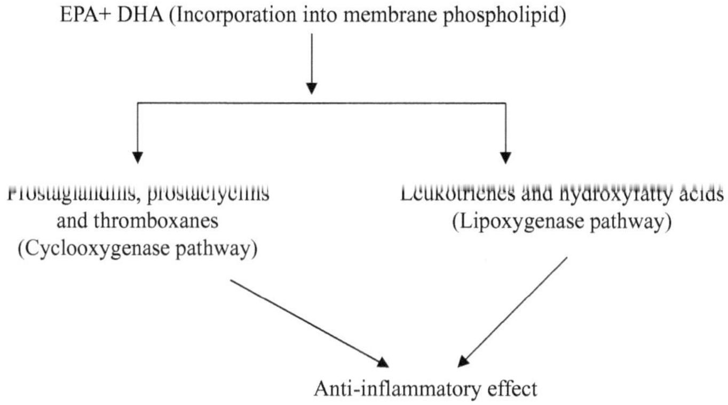

Figure 5.6 Anti-inflammatory mechanism of eicosanoids

derived from linolenic acid have anti-inflammatory responses (Figure 5.6) (James et al., 2000; Barcelo-Coblijn and Murphy, 2009; Kaur et al., 2014).

Factors such as bile secretion, intestinal pH and microorganisms may bring drastic environmental change along the gastrointestinal tract, and this probably affects the absorption of essential fatty acids. Many epidemiological studies have reported that intake of fatty acids with cholesterol supplementation stimulates fatty acid absorption, which may be because this favors the increase in micelle formation and binding of gut content for the intestine brush border, by which absorption takes place (Dupasquier et al., 2007). Shahidi and Ambigaipalan (2018) reported in their review that both EPA and DHA are firstly hydrolyzed into free fatty acids by pancreatic carboxylic acid ester lipase. They are absorbed by enterocytes. For uptake of FA by enterocytes, a membrane-bound fatty acid binding protein is involved. As for fatty acids, there might be fatty acid transport proteins as a major factor affecting their absorption. The enterocytes are then incorporated into chylomicrons, which ultimately secrete into lymph. Efficient intraluminal absorption takes place in the jejunum, and the ileum allows only some of the ingested FA to pass into feces.

5.2.2 Mechanism of Essential Fatty Acids in Health

5.2.2.1 Inflammation

It is evident from many studies that the anti-inflammatory activity of DHA is partially attributable to its effect a) on cell membrane reorganization and b) on specific receptors, as DHA modulates inflammatory gene expression through intracellular receptors (GPR120 and PPARϒ) by binding with them (Laye et al., 2018). In addition to providing health benefits, both ARA (arachidonic acid) and EPA serve as precursors for the formation of eicosanoids. Evidence from much research has proven that ARA derived from omega-6 fatty acids has a crucial role in mediating the inflammatory response; however, omega-3 fatty acids are responsible for inflammation. EPA is used as substrate for the cyclooxygenase and lipoxygenase enzymes to produce eicosanoid and in combination with DHA inhibits ARA metabolism in the cell membrane by decreasing the expression of the cox2 gene (responsible for inflammation) (Cadler, 2013). Omega-3 has the potential to compete with enzymes that convert AA into eicosanoids like PGs, TXs and LTs, which are responsible for inflammation. However, eicosanoids derived from EPA (PGE3 and LTB5) are biologically less active than ARA-derived eicosanoids (PGs, TXs and LTs responsible for inflammation) and they have anti-inflammatory properties (Engler, 2017). Omega-3 fatty acids displace ARA from cell membranes' phospholipids and reduce the inflammatory and pain-promoting properties provided by omega-6 fatty acids. As explained by Cadler (2013), EPA and DHA are capable to inhibit a number of aspects of inflammation including production of eicosanoids like prostaglandins and leukotrienes from ADA, responsible for production of inflammatory cytokines. Again in 2017, Cadler (2017) reported that EPA and DHA give rise to anti-inflammatory and inflammation-resolving mediators called resolvins, protectins and maresins.

5.2.2.2 Cardiovascular Disease

Diet patterns, genetic factors and the modern lifestyle have become the main reasons for coronary heart disease (CHD), which is responsible for a great deal of the unhealthy state of mind and mortality in the world. It was Sinclair, a British physiologist, who firstly hypothesized the protective role of omega-3 fatty acids in preventing CHD and their preventive role was first demonstrated in studies of Greenland Inuits, who have a higher consumption of fish oil in lowering CVD. Further, many researchers performed pioneering studies among Greenland Eskimos and suggested that high intakes of EPA and DHA can protect them against CHD. In 2012, a major clinical study was conducted by Kaur et al. (2014) with over a thousand patients, providing them 1 g/day of omega-3 fatty acids; this study indicated great reductions of 20%, 30% and 45% in death, cardiovascular disease and cardiac failure, respectively. The prevention mechanisms of omega-3 fatty acids involve reductions in circulating triacylglycerol levels, platelet activation and the expression of vascular adhesion molecules (Vanschoonbeek et al., 2003). It was further understood that triacylglycerols enriched with omega-3 fatty acids are less dependent on lipoprotein lipase activity and apolipoprotein E-dependent pathways for blood removal and are removed faster from plasma (Qi et al., 2002).

5.2.2.3 Blood cholesterol

Epidemiological and genetic studies suggested that hypercholesterolemia and cardiovascular diseases are correlated with each other. Zibaeenezhad et al. (2017) reported that a 1-mg/dl increase in the low-density lipoprotein level may increase (2–3%) risk for CVDs and that higher elevations may be associated with further increases in risk. Prescription formulations of EPA and DHA are capable of improving the lipid profile of normal lipidemic individuals (Egert et al., 2009). ALA has lipid-lowering ability and anti-atherosclerotic properties along with low plasma LDL levels. The cholesterol-lowering effects of ALA agreed with results from many researchers who have used flaxseed oil, canola oil, walnut oil and olive oil in individuals with normal or high cholesterol levels.

5.2.2.4 Hypertension

Hypertension affects up to 20% of populations and contributes a major burden in cardiovascular disease. Through its antioxidant properties, omega-3 fatty acids may improve vascular function and reduce arterial stiffness and atherosclerotic damage (Colussi et al., 2016). Evidence has suggested that EPA and DHA have multiple health benefits, i.e., reducing systolic and diastolic blood pressure (BP), lipid metabolism, heart rate and platelet function. A study conducted by deCamargo et al. (2015) on lifestyle modification reported on the BP-lowering effects of omega-3. Hoshi et al. (2013) reported that a dose of 3 g/d EPA and DHA led to a reduction in BP without altering the breathing rate.

5.2.2.5 Cancer

Throughout the world, cancer is a major disease burden and conventional therapies are not as beneficial as desired because of the drug resistance of cancer cells and these therapies' toxicity to normal growing cells (D'Eliseo and Velotti, 2016). During the past decade, understanding of the mechanisms of carcinogenesis, cancer progression and resistance to treatment has been achieved. Dietary fat has been one of the most frequently investigated dietary risk factors for prostate cancer, and most findings support no association for total fat. EPA and DHA may exert anti-cancer activity by inducing apoptotic cell death in cancer cells either alone or with conventional therapies. High levels of DHA, the most abundant ω-3 PUFA in erythrocyte membranes, were associated with a reduced risk of breast cancer (Fabian et al., 2015). Wendel and Heller (2009) in their review concluded that omega-3 fatty acid supplements lessen growth and encourage apoptosis in cancer cell lines procured from prostate, breast, pancreatic and colonic cancer.

5.3 BIOAVAILABILITY

The bioavailability of omega-3 is influenced by the form in which it exists: ethyl ester, triglycerols, phospholipids and free fatty acids, along with diet enriched with the ester form of omega-3 fatty acid, have lower bioavailabilities than FFA (Schuchardt et al., 2011). As reported by researchers, for whole-body utilization of ALA, the β-oxidation pathway is more responsible. ALA is reported to have low bioavailability due to its high rate of oxidation (Libinaki and Gavin, 2017) with 85% of consumed ALA disappearing. In contrast, DHA, being a poor substrate for β-oxidation pathways, is comparatively more bioavailable. The bioavailabilities of fatty acids are affected by their position on the glycerol molecule of the triglycerol. sn-2 is the most favored position and results in easier absorption and ultimately greater bioavailability (Tengku-Rozaina and Birch, 2013). It is reported that 58.9% of the total amount of ALA ingested disappeared, 0.4% was excreted in feces and 21.2% accumulated as ALA.

5.4 CONCLUSIONS

Essential fatty acids have become the focus of scientific research and are starting to become widely accepted as a part of modern nutrition because they contribute beneficial effects on metabolism. Dietary patterns including consumption of EFA are positively correlated with reducing morbidity and mortality from CVD and anti-inflammatory and anticarcinogenic potentials. Lots of research has been conducted on omega-3/omega-6 health benefits, but still there is a lack of knowledge of appropriate doses of omega-3 and omega-6 and the correct omega-3/omega-6 ratio. A further understanding of the efficiency and long-term safety of ALA is also required.

REFERENCES

Balk, E.M. and Lichtenstein, A.H. 2017. Omega-3 fatty acids and cardiovascular disease: Summary of the 2016 Agency of Healthcare Research and Quality Evidence Review. *Nutrients* 9(8): 865.

Barceló-Coblijn, G. and Murphy, E.J. 2009. Alpha-linolenic acid and its conversion to longer chain n-3 fatty acids: Benefits for human health and a role in maintaining tissue n-3 fatty acid levels. *Progress in Lipid Research* 48(6): 355–374.

Calder, P.C. 2017. Omega-3 fatty acids and inflammatory processes: From molecules to man. *Biochemical Society Transactions* BST20160474: 1105–1115.

Calder, P.C. 2013. Omega-3 polyunsaturated fatty acids and inflammatory processes: Nutrition or pharmacology? *British Journal of Clinical Pharmacology* 75(3): 645–662.

Carlier, H., Bernard, A. and Caselli, C. 1991. Digestion and absorption of polyunsaturated fatty acids. *Reproduction, Nutrition, Development* 31(5): 475–500.

Christensen, M.S., Müllertz, A. and Høy, C.E. 1995. Absorption of triglycerides with defined or random structure by rats with biliary and pancreatic diversion. *Lipids* 30(6): 521–526.

Colussi, G., Catena, C., Novello, M. and Sechi, L.A. 2016. Omega-3 polyunsaturated fatty acids in blood pressure control and essential hypertension. In: Salazar-Sanchez, L. (ed.), *Update on Essential Hypertension*. London: InTech.

deCamargo Talon, L., deOliveira, E.P., Moreto, F., Portero-McLellan, K.C. and Burini, R.C. 2015. Omega-3 fatty acids supplementation decreases metabolic syndrome prevalence after lifestyle modification program. *Journal of Functional Foods* 19: 922–928.

D'Eliseo, D. and Velotti, F. 2016. Omega-3 fatty acids and cancer cell cytotoxicity: Implications for multi-targeted cancer therapy. *Journal of Clinical Medicine* 5(2): 15.

Dupasquier, C.M.C., Dibrov, E., Kneesh, A.L., Cheung, P.K.M., Lee, K.G.Y., Alexander, H.K., Yeganeh, B.K., Moghadasian, M.H. and Pierce, G.N. 2007. Dietary flaxseed inhibits atherosclerosis in the LDL receptor-deficient mouse in part through antiproliferative and anti-inflammatory actions. *American Journal of Physiology: Heart and Circulatory Physiology* 293(4): H2394–H2402.

Engler, M.M. 2017. Role of dietary omega-3 fatty acids in hypertension. *Annals of Nursing and Practice* 4(1): 1077.

Egert, S., Kannenberg, F., Somoza, V., Erbersdobler, H.F. and Wahrburg, U. 2009. Dietary α-linolenic acid, EPA, and DHA have differential effects on LDL fatty acid composition. *The Journal of Nutrition* 139(5): 861–868.

Fabian, C.J., Kimler, B.F. and Hursting, S.D. 2015. Omega-3 fatty acids for breast cancer prevention and survivorship. *Breast Cancer Research: BCR* 17(1): 62.

Hoshi, T., Wissuwa, B., Tian, Y., Tajima, N., Xu, R., Bauer, M. and Hou, S. 2013. Omega-3 fatty acids lower blood pressure by directly activating large-conductance Ca2+-dependent K+ channels. *Proceedings of the National Academy of Sciences* 110(12): 4816–4821.

James, M.J., Gibson, R.A. and Cleland, L.G. 2000. Dietary polyunsaturated fatty acids and inflammatory mediator production. *The American Journal of Clinical Nutrition* 71(1): 343s–348s.

Kaur, N., Chugh, V. and Gupta, A.K. 2014. Essential fatty acids as functional components of foods: A review. *Journal of Food Science and Technology* 51(10): 2289–2303.

Laye, S., Nadjar, A., Joffre, C. and Bazinet, R.P. 2018. Anti-inflammatory effects of omega-3 fatty acids in the brain: Physiological mechanisms and relevance to pharmacology. *Pharmacological Reviews* 70(1): 12–38.
Libinaki, R. and Gavin, P.D. 2017. Changes in bioavailability of omega-3 (dha) through alpha-tocopheryl phosphate mixture after oral administration in rats. *Nutrients* 9(9): 1042.
Moon, Y.A., Hammer, R.E. and Horton, J.D. 2009. Deletion of ELOVL5 leads to fatty liver through activation of SREBP-1c in mice. *Journal of Lipid Research* 50(3): 412–423.
Patterson, E., Wall, R., Fitzgerald, G.F., Ross, R.P. and Stanton, C. 2012. Health implications of high dietary omega-6 polyunsaturated fatty acids. *Journal of Nutrition and Metabolism* 2012: 1–16.
Qi, K., Seo, T., Al-Haideri, M., Worgall, T.S., Vogel, T., Carpentier, Y.A. and Deckelbaum, R.J. 2002. Omega-3 triglycerides modify blood clearance and tissue targeting pathways of lipid emulsions. *Biochemistry* 41(9): 3119–3127.
Rustan, A.C. and Drevon, C.A. 2001. Fatty acids: Structures and properties. *e LS*.
Schuchardt, J.P., Schneider, I., Meyer, H., Neubronner, J., VonSchacky, C. and Hahn, A. 2011. Incorporation of EPA and DHA into plasma phospholipids in response to different omega-3 fatty acid formulations--a comparative bioavailability study of fish oil vs. krill oil. *Lipids in Health and Disease* 10(1): 145.
Shahidi, F. and Ambigaipalan, P. 2018. Omega-3 polyunsaturated fatty acids and their health benefits. *Annual Review of Food Science and Technology* 9: 345–381.
Simopoulos, A.P. 2016. An increase in the omega-6/omega-3 fatty acid ratio increases the risk for obesity. *Nutrients* 8(3): 128.
Sprecher, H. 2000. Metabolism of highly unsaturated n-3 and n-6 fatty acids. *Biochimica et Biophysica Acta (BBA)-Molecular and Cell Biology of Lipids* 1486(2–3): 219–231.
Stoffel, W., Holz, B., Jenke, B., Binczek, E., Günter, R.H., Kiss, C., Karakesisoglou, I., Thevis, M., Weber, A.A., Arnhold, S. and Addicks, K. 2008. Δ6-Desaturase (FADS2) deficiency unveils the role of ω3-and ω6-polyunsaturated fatty acids. *The EMBO Journal* 27(17): 2281–2292.
Tengku-Rozaina, T.M. and Birch, E.J. 2013. Enrichment of omega-3 fatty acids of refined hoki oil. *Journal of the American Oil Chemists' Society* 90(8): 1111–1119.
Vanschoonbeek, K., deMaat, M.P. and Heemskerk, J.W. 2003. Fish oil consumption and reduction of arterial disease. *The Journal of Nutrition* 133(3): 657–660.
Wendel, M. and Heller, A.R. 2009. Anticancer actions of fatty acids-current state and future perspectives. *Anti-Cancer Agents in Medicinal Chemistry* 9(4): 457–470.
Zibaeenezhad, M.J., Ghavipisheh, M., Attar, A. and Aslani, A. 2017. Comparison of the effect of omega-3 supplements and fresh fish on lipid profile: a randomized, open-labeled trial. *Nutrition & Diabetes* 7(12): 1–8.

6

Effect of Thermal and Non-Thermal Processing on EFA

Priyanka Kajla and Vandana Chaudhary

Contents

6.1 INTRODUCTION

Essential fatty acids (EFA) are required for perfect health maintenance, but cannot be synthesized in the body; therefore, these must be obtained from different types of food sources. The human body is not able to synthesize essential fatty acids due to the absence of desaturase and enlongase enzymes, which are required during their synthesis (de Lorgeril et al., 2001). Basically, polyunsaturated fatty acids are classified as omega-6 and omega-3. The parent omega-6 fatty acid is referred as to linoleic acid while the parent omega-3 fatty acid is termed as alpha-linolenic acid (ALA). Both omega-3 and 6 fatty acids are considered as essential fatty acids. ALA serves as the precursor for the synthesis of long-chain fatty acids—EPA (eicosapentaenoic acid) and DHA (docosahexaenoic acid) (Kajla et al., 2015). The health benefits of omega-3 and -6 fatty acids are well established. A large number of clinical studies have recognized the tremendous potential of omega-3 polyunsaturated fatty acids against inflammatory mediators like prostaglandins E2, leukotriene B4, TNF-α, interleukin and cytokines. These clinical studies revealed that n-3 polyunsaturated fatty acids are helpful in prevention of coronary heart diseases, atherosclerosis, rheumatoid arthritis and asthma (Arend and Dayer, 1995; Kajla et al., 2015; Punia et al., 2019). Daily intake of 3 g EPA and DHA for more than 12 weeks was found to be effective in reducing the inflammation of rheumatoid arthritis (Kremer, 2000). One of the major challenges in handling/processing of polyunsaturated fatty acids is their high sensitivity to oxidation. Inclination of lipid oxidation in food depends on the kind of fatty acids, the presence of oxygen, temperature, and light and the use of different processing methods. The presence of redox active metals such as Fe^{2+} and Cu^{2+}, the presence of heme proteins, elevated concentrations of oxygen and elevated temperatures can further hasten the oxidative reactions of polyunsaturated fatty acids (Berlitz et al., 2009). Light and heat trigger the reaction of long-chain polyunsaturated fatty acids in the autoxidation chain by increasing the activation energy required for the oxidation reaction and generating the free radicals necessary for chain propagation, which is thermodynamically difficult without light and heat and despite the presence of air (Hadaruga et al., 2016). Oxidation of lipids often leads to dilemmas during the processing and storage, particularly of long-chain unsaturated fatty acids. Highly unsaturated fatty acids are unstable particularly during thermal processing, as heat induces a number of chemical transformations: oxidation, polymerization, cyclization, geometrical isomerization and/or double bond migration. Among the degradation products formed owing to thermal processing are oligomers, cyclic fatty acid monomers and *trans* fatty acid isomers (Zhang et al, 2012; Ludger, 2014). Production of such compounds reduces the nutritive value of polyunsaturated fatty acid–enriched foods. Additionally, these oxidative products of polyunsaturated fatty acids are reported to be directly related to the occurrence of several health problems.

6.2 THERMAL PROCESSING OF ESSENTIAL FATTY ACIDS

Heat-induced oxidative deterioration of the polyunsaturated fatty acids ensues primarily due to prolonged heating exposure, which produces a mixture of volatile secondary oxidation products causing objectionable off-flavors in the foods. Basically, thermal processing of long-chain unsaturated fatty acids induces the quality alterations in foods enriched with these fatty acids in two ways (Let et al., 2005). *Firstly*, it adversely affects the flavor of foods, which might be attributed to the oxidative products particularly aldehydes and ketones. For example, butter enriched with polyunsaturated fatty acids/conjugated linoleic acid produces a wide range of oxidative products including 3-methyl-1H-indole, pentanal, heptanal and butanoic acid producing mothball-like, fatty, green and cheesy off-flavors, respectively (Mallia et al., 2009). *Secondly*, the degradation of polyunsaturated fatty acids results in a loss of nutritional value by causing the obliteration of omega-3 and 6 fatty acids as well as heat-susceptible vitamins A, D, E and K, thus lowering the calorific value of the foods (Arab-Tehrany et al., 2012).

It is well established that thermal decomposition of long-chain fatty acids requires free radical production to trigger reactions. The initial step follows the decomposition of primary metabolite peroxides, by means of heat or catalytic reactions. Production of an adequate amount of free radicals propagates the chain reactions by abstracting hydrogen from complex molecules. The mechanisms and process of formation of decomposition products are very complicated, because not only does scission of the C–C double bond take place but various decomposition reactions ensue simultaneously. As a result of these chain reactions mediated by free radical reactions, a variety of degradation products are produced, viz., *trans* and free fatty acids, polymerized glycerides, short-chain saturated fatty acids, conjugated fatty acids, epoxy, ketonic acids and cyclic fatty acids as primary metabolites. Further aldehydes, hydroxy aldehydes and hydrocarbons are produced as secondary metabolites. The high temperature developed during cooking/processing of long-chain fatty acids impedes free radical formation or generates singlet oxygen species that first trigger/accelerate the oxidation process leading to formation of peroxide radicals, which upon further oxidation, form different compounds such as carbonyl compounds (–CHO/–C=O), alcohols, esters and short-chain hydrocarbons, contributing to the development of off-flavors as well as reductions in the nutritional and organoleptic quality of PUFA-enriched foods. In addition to temperature, polyunsaturated fatty acids are also prone to oxidative deterioration induced by the presence of light, superoxide radical anion-generating enzymes and metal ions, which are known to be participants in the chain initiation of auto-oxidation reactions of long-chain fatty acids. Volatile organic acids are also reported to be produced during the oxidation of fats/lipids (Huang and Sathviel, 2008; Hadarunga et al., 2016).

One of the major challenges in processing of long-chain fatty acids is their high vulnerability to the presence of oxygen and elevated temperatures. The pace of oxidative damage of long-chain essential fatty acids increases drastically with increases in temperature. It has been reported that the relative oxidation rate of polyunsaturated fatty acids particularly of alpha-linolenic acid is approximately 2500 times higher than that of saturated stearic acid (Berlitz et al., 2009). A few studies demonstrated the higher vulnerability of polyunsaturated fatty acids as compared with that of saturated fatty acids. The stability of the lipids toward the oxidation process decreases with the increase in the extent of unsaturation. Oleic acid is estimated to be approximately 25-fold more stable than linoleic acid and 75-fold more stable than linolenic acid (Hrncirik, 2010).

It has been reported that with the decrease in linoleic acid levels, the increase in the oxidative stability of oils is noticeable, resulting in an improvement of organoleptic characteristics, especially in terms of reduced bitterness. A study of the thermo-oxidative stability of different edible oils rich in long-chain polyunsaturated fatty acids revealed that these oils have higher peroxide, anisidine and totox values, indicating the severity of oxidation during processing and storage (Abdulkarim et al., 2005).

During frying with different oils, there was a downtrend in the linoleic acid content, while saturated fatty acids, especially palmitic acid, stearic acid and oleic acid, were found to be increased with frying time. This increase in the content of saturated fatty acids during frying or high-temperature processing is attributed to their strong oxidative stability over polyunsaturated fatty acids. At low temperatures as well, the omega-3 and 6 fatty acids are extremely susceptible to oxidative processes. The levels of fatty acids, in particular EPA and DHA, drop down by about 66% at a processing temperature of 50°C, and are further reduced by 75% with increment in processing temperature of 150°C (Sharoba and Ramadan, 2012).

Salmon fish, which is rich in essential fatty acids, was subjected to different thermal processing methods, namely boiling, pan-frying and oven-baking, to examine the effects on the levels of long-chain omega-3 and 6 fatty acids and the extent of oxidation of these fatty acids (Leung et al., 2018). It was reported that none of the above processing methods significantly reduced the content of either of the essential fatty acids in the salmon samples, indicating no quality alterations after thermal processing. Thermal exposure of long-chain fatty acids induces different oxidative degradation products that might act beneficial or harmful to human health.

Of thermal processing methods studied, frying and baking induced maximum oxidative degradation of polyunsaturated fatty acids in variously processed salmon samples. The thermo-oxidative degradation of the unsaturated fatty acids in processed salmon varied due to the difference in times of exposure to different temperatures employed for different processing methods. Boiling was performed at 100°C by immersing salmon in hot water, and transfer of heat by hot water is very efficient because of continuous collisions among the lipid molecules by the mode of convection. The short duration of processing along with minimal processing temperature might explain the development of low levels of peroxide and carbonyl compound formation in boiled salmon samples. In the case of pan-frying (210°C), heat transfer takes place through the conduction mode, which led to rapid surface dehydration, thus inducing browning reactions, and the exposure of the salmon surface to a high temperature along with the oxygen and moisture triggers the formation of free radicals. Therefore, it is evident from the investigation that maximum thermo-oxidative degradation of long polyunsaturated fatty acids was noticed in fried samples in comparison to boiled salmon samples, leading to nutritional loss. However, even if the baking temperature and exposure time are higher, the rate of lipid peroxidation is comparatively less than pan-frying. This might be because the heat transfer in baking takes place by air convection and radiation, and therefore, the energy transfer is low. It, therefore, requires lengthy contact/exposure of the food to heat in the oven, but the temperature inside the food is much less than the temperature outside (i.e., the temperature of the oven), which leads to comparatively lesser thermo-oxidative degradation of heat-sensitive vital fatty acids and other constituents (Leung et al., 2018).

The presence of numerous double bonds in long-chain polyunsaturated fatty acids, especially EPA and DHA, makes them highly prone to thermo-oxidative degradation at the elevated temperatures produced by the above-mentioned processing techniques. A growing number of research studies have shown that oxidative metabolites produced by the oxidation of polyunsaturated omega-3 fatty acids in low concentrations may not influence human health. F3-isoprostanes (F3-IsoPs), oxidative products of EPA, are reported to have anti-inflammatory and antithrombotic properties because of their capability to reduce the output of arachidonic acid–derived pro-inflammatory compounds viz., prostaglandins and thromboxanes by competing for COX (cyclo-oxygenase) enzyme activity. However, the oxidation of DHA is known to produce a secondary oxidation product known as 4-hydroxy-2-hexenal (4-HHE), which is reported to have neurotoxic properties, by increasing reactive oxygen species activity and decreasing the concentrations of antioxidant enzymes and glutathione (GSH). Recently, studies have demonstrated that a low concentration of this compound is good for cardiovascular health by activating Nrf2 in vascular cells (Larsson et al., 2016 and Leung et al., 2018).

In line with the above studies, humpback salmon fish was boiled, fried or roasted and it was observed that out of these processing methods frying caused a significant reduction of omega-3 and 6 fatty acids. A decline in the concentration of essential fatty acids may be attributed to the longer period of frying (Gladyshev et al., 2006).

On the contrary, a few research investigations stated that the type of oil used for frying, but not the extent of frying, affects the quality and quantity of the heat-susceptible polyunsaturated fatty acids. Pan-frying of omega-3–enriched fish was carried out in corn, peanut, canola and sunflower oil. It was found that frying using canola and corn oil caused nonsignificant reductions in EPA and DHA levels, and frying in peanut oil led to the reduction in EPA levels only, while sunflower oil–fried fish had significantly reduced levels of long-chain omega-3 fatty acids. These trials proposed that the lipids of the fish leach into the cooking oil during the frying process and that the peroxide radicals produced in the hot cooking oil caused degradation of fatty acids and thus reduced the levels of essential polyunsaturated fatty acids in the processed fish (Sioen et al., 2006 and Flaskerud et al., 2017).

Processing of skipjack tuna (*Katsuwonus pelamis*) was done using different methods viz., cooking in boiling water bath (100°C), microwave cooking, shallow frying, baking and canning. This research showed that boiling and baking are the finest processing techniques to preserve the polyunsaturated fatty acids, especially omega-3 fatty acids, in comparison to frying and canning processing. It was noticed that cooking tuna in a boiling water bath led to nonsignificant degradation of long-chain polyunsaturated fatty acids, in particular omega-3 fatty acids, while approximately 25% of EPA and 55% of DHA were destroyed in microwave heating. Further, these fatty acids were completely destroyed by canning, while a 70% to 85% reduction was noticed in omega-3 fatty acids during the process of frying (Stephen et al., 2010).

Furan (C_4H_4O) is a carcinogenic compound formed during thermal processing of highly unsaturated fatty acids. The *trans* isomers of conjugated linolenic acid are reported to be highly prone to oxidation to form furan fatty acids. Significant amounts of furan fatty acids were formed during oxidative degradation of omega-3 and 6 fatty acids during thermal processing. A significant amount of furan formation takes place during thermal treatment of alpha-linolenic acid. Even other unsaturated fatty acids form furan during thermal processing but in very small amounts (Fan, 2015). Becalski and Seaman (2005) verified that linolenic acid produced more furan as compared to linoleic acid when exposed to 118°C for 30 minutes.

Geometrical isomerization occurs during deodorization of fish, which is generally conducted at temperatures in the range of 180–270°C under vacuum in the presence of steam. By geometrical isomerization, 32 and 64 isomers of eicosapentaenoic acid and docosahexaenoic acid can be formed, respectively. The geometrical isomerization of PUFA is not a negligible phenomenon, even if other reactions (polymerization, oxidation) are quantitatively more important in edible fats and oils. Fatty acid compositions of the fish oils show that deodorization has an important effect on the geometrical isomerization of EPA and DHA. Evolution in *trans* isomers of EPA and DHA occurs as a function of deodorization temperature. Due to the possible overlapping of *trans* fatty acids with other degradation products, e.g., CFAM, the reported values for *trans* fatty acids might be slightly overestimated. Only minor amounts of EPA and DHA *trans* isomers were formed during deodorization at 180°C. However, when the oil was deodorized at a temperature of 220°C, about 7.7 and 34.1 mg/g of EPA and DHA geometrical isomers were formed respectively. Deodorization at 250°C was so detrimental to PUFA that only 0.9% EPA and 2.2% DHA remained in the fish oil after such a treatment. Furthermore, 20.0 and 55.6 mg/g of geometrical isomers of eicosapentaenoic acid and docosahexaenoic acid, respectively, were detected in the deodorized oil at 250°C (Fournier et al., 2006).

It can therefore be concluded from the above discussion that the stability of a particular oil during thermal processing depends on the types and amounts of fatty acids present and the duration of exposure. Both essential polyunsaturated fatty acids, omega-3 and 6, are highly susceptible substrates during thermal processing, which might result in their potential loss quantitatively as well as qualitatively.

6.3 NON-THERMAL PROCESSING OF ESSENTIAL FATTY ACIDS

Essential fatty acids, particularly omega-3 fatty acids, have been shown to provide protection against inflammatory disorders, type II diabetes, joint disorders, cancer, neurodegenerative disorders, schizophrenia, autism, cardiovascular diseases and Alzheimer's disease (Calder, 2013). The inclusion of these essential fatty acids into the diet through the processing of foods is a very challenging task because of their acute sensitivity toward oxidation, which is attributed to the higher degree of unsaturation. The prevention of the oxidation process of polyunsaturated fatty acids has been a challenge in recent decades, especially for highly unstable essential fatty acids. Efforts are being made consistently in this context to use antioxidants along with synergistically compatible compounds to increase their efficacy, but these have not been very successful (Ghnimi et al., 2017).

6.3.1 Micro- and Nano-Encapsulation

Jacobsen (2015) reviewed the different approaches used to stabilize essential fatty acid–enriched foods and addressed the importance of optimizing the parameters for antioxidant supplementation and processing. It has been established that the use of compound antioxidants with multifunctional attributes, such as free radical and reactive oxygen species entrapping, metal chelating and capability of antioxidant regeneration, might increase the efficacy of the antioxidants with a single specific property. In addition to the addition of antioxidants, a further approach for protecting the nutritionally important essential fatty acids from oxidation was encapsulation. This technique is also used to facilitate the incorporation of encapsulated essential fatty acids in various processed food products (Barrow et al., 2013).

The term "encapsulation" refers to a technique in which a compound that needs to be protected is embedded in a matrix of proteins or polysaccharides using different processing techniques like spray-drying, freeze-drying, etc. Spray-drying is a widely practiced technique for producing microencapsulated food products because spray-driers are easily accessible, and it has lower production costs than other methods. The electrospraying technique offers spraying as compared to spray-drying and also utilizes a huge amount of hot gas for drying, thus adding to the production costs (Gomez-Mascaraque and Lopez-Rubio, 2016).

With respect to omega-3 and -6 fatty acid stability, the encapsulation of these heat-sensitive fatty acids depends on numerous constraints viz., the quality of the oil, oil distribution within the particle, the type of antioxidants, the combination of antioxidants and synergist compounds, moisture, water activity, the particle size/density and the composition of the microcapsule (Gharsallaoui et al., 2007).

Encapsulation technologies play a crucial role in preventing thermo-oxidation and thus extending the shelf stability of essential fatty acids, offering realistic options for oxidative stabilization of omega-3 PUFA-enriched foods. While encapsulating the heat-labile essential fatty acids, it is imperative to manage the oil distribution over the microcapsule surface as well as its contact/exposure to the atmosphere and heat during processing that could otherwise cause adverse quality changes in the encapsulated fatty acid (Velasco et al., 2006). In addition, the concept of multilayered encapsulation technology is also utilized to further enhance the efficacy of the capsules. The incorporation of protein hydrolyzates, gums and carbohydrate polymers can strengthen the capsules as well as contribute to enhanced stability of the encapsulated polyunsaturated fatty acids (Shaw et al., 2007).

6.3.2 Antioxidants, Surfactants and Emulsifiers

The incorporation of antioxidants in the fatty acid encapsulates to form microcapsules with enhanced shelf stability is also trending. The incorporated antioxidants have free radical– and oxygen-scavenging properties along with the capability to form chelates or complexes with metals, which aids in the protection of microcapsules from oxidation (Ghnimi et al., 2017).

Another approach for providing maximum stability to the micro- and nano-encapsulates containing functional fatty acids is by the addition of emulsifiers and surfactants. The emulsifiers to be added in encapsulates should have properties for imparting stability to nano- and microemulsions/capsules having essential fatty acids (Walker et al., 2015). Nonionic, anionic, cationic and zwitterionic surfactants are some of the surfactants added to capsules containing omega-3 fatty acids, while phospholipids, proteins, polysaccharides, saponins and glycolipids are generally used emulsifiers. The incorporation of surfactants and emulsifiers not only imparts oxidative stability to long-chain fatty acids but also contributes to the physical strength of micro- and nano-emulsions (Liu et al., 2016; He et al., 2017). The packaging protocols are also reported to have a significant effect on the storage stability of EPA and DHA microcapsules that can be packaged in the presence of inert nitrogen gas atmosphere in order to extend the shelf life of the product (Ghnimi et al., 2017).

6.3.3 Hydrogels

Novel hydrogels were fabricated using cress seed mucilage (CSM) and chitosan (CS) for the encapsulation of fish oil, and their different physicochemical features were studied. Hydrogels were effective in protection of fish oil against oxidation. Thermal analysis revealed that combination of CSM (with thermal stability at 315°C) and CS (with thermal stability at 338°C) led to thermal stability enhancement up to 355°C of produced hydrogel. The CSM/CS hydrogel was effective for preventing fish oil oxidation after 14 days. Pronounced oxidative stability of optimum CSM/CS hydrogels was related to its capability to entrap high amounts of fish oil (Kavousi et al., 2017).

6.3.4 Cold Plasma

The technique of cold plasma (CP) has aroused significant interest on the part of food processors as an alternative to thermal processing in the area of food processing and preservation. Plasma is a neutral ionized gas with a net neutral charge consisting of atoms, ions and free electrons in their ground or energized states. Cold plasma processing is a novel technology usually known for decontamination and surface treatment, but it is also reported to significantly affect the different characteristics of foods during processing and storage (Pankaj et al., 2014). Limited literature is available from studies concerning the effect of cold plasma processing on polyunsaturated fatty acids. However, based on scientific investigations, treatment duration and plasma used are the crucial factors that affect the oxidation of unsaturated lipids. Nonsignificant results in terms of oxidation were observed in the fatty acid profiles of fresh and frozen pork after treatment with plasma gases. However, other studies reported that lipid oxidation was enhanced in unprocessed/fresh pork and beef after giving a cold plasma gas treatment for 10 minutes. Similarly, Albertos et al. (2017) observed significant changes in oxidation rates of fatty acids of mackerel fillets following cold plasma treatment. Cold plasma–treated fillets showed marked increases in peroxide value, indicating increased oxidative damage. A noticeable decrement was reported in levels of oleic acid and eicosapentaenoic acid after plasma treatments. Yepez and Keener

(2016) have recently reported a novel use of cold plasma processing technology. The hydrogen plasma technology can be utilized for the preparation of hydrogenated fats without the formation of *trans* fatty acids, which otherwise is of major concern. Along with this, cold plasma technology is nowadays a preferred technique over the traditional hydrogenation process, as it is a simple technique that can be applied at room temperature, without utilization of any catalyst under atmospheric pressure.

6.3.5 Pulsed Electric Field

Pulsed electric fields (PEFs) are also being applied for processing and preservation of polyunsaturated fatty acids. Various research findings have demonstrated that PEFs bring out changes in the structural and textural properties of meat lipids, thereby improving the functionality of lipids, which is desirable for new product development (McDonnell et al., 2014). On the other hand, PEF is also reported to induce undesirable changes in meat. Raw meat stored at low temperature is usually oxidatively stable but basic processing steps viz., mincing, cooking, emulsification etc., cause disruption of muscle cell membranes, thus exposing the oxidation catalysts (heme/nonheme iron) and polyunsaturated fatty acid to induce the oxidation process (Arroyo et al., 2015). The oxidation of polyunsaturated fatty acids generates free radicals and various primary and secondary oxidative metabolites. These metabolic products further degrade into acids, alcohols, aldehydes and ketones, which produces oxidized, rancid and off-flavors (Byrne, 2000). Thus, PEF treatment could cause damage to the cell membranes, promoting the inception of lipid oxidation, thereby deteriorating the nutritive value and hence the quality of meat lipids. However, another study revealed that the PEF treatment did not affect the linolenic and oleic acid contents in grape juice.

6.4 CONCLUSION

The long-chain polyunsaturated fatty acids are the most vulnerable constituents for oxidation and polymerization during thermal treatment of almost all vegetable oils. Therefore, these fatty acids can serve as one of the best indicators of quality changes in edible vegetable oils during thermal processing. Based on these recommendations, polyunsaturated fatty acids can be appropriately processed so that essential fatty acids along with fat-soluble vitamins might be retained to maximum levels, as this will lead to significant increases in the dietary intake of these vital nutrients as well as ensure the safety of polyunsaturated fatty acid–enriched processed foods. Non-thermal technologies have recently emerged as alternatives over thermal treatment for processing polyunsaturated fatty acid–enriched foods. More studies need to be conducted to explore the potential of non-thermal technologies for maximum retention of the essential fatty acids. However, non-thermal technologies have certain limitations in ensuring the alterations in organoleptic, physical and chemical properties of foods. It is now a challenge to food technologists and researchers to tailor non-thermal technologies to focus on minimizing quality changes and maximizing retention and stability not only of essential fatty acids but also of other vital nutrients.

REFERENCES

Abdulkarim, S.M., Long, K., Lai, O.M., Muhammad, S.K.S. and Ghazali, H.M. 2005. Some physico-chemical properties of *Moringa oleifera* seed oil extracted using solvent and aqueous enzymatic methods. *Food Chemistry* 93(2):253–263.

Albertos, I., Martin-Diana, A., Cullen, P., Tiwari, B., Ojha, S., Bourke, P., Álvarez, C. and Rico, D. 2017. Effects of dielectric barrier discharge (DBD) generated plasma on microbial reduction and quality parameters of fresh mackerel (*Scomber scombrus*) fillets. *Innovative Food Science and Emerging Technologies* 44:117–122.

Arab-Tehrany, E., Jacquot, M., Gaiani, C., Imran, M., Desobry, S. and Linder, M. 2012. Beneficial effects and oxidative stability of omega-3 long-chain polyunsaturated fatty acids. *Trends in Food Science and Technology* 25(1):24–33.

Arend, W.P. and Dayer, J.M. 1995. Inhibition of the production and effects of interleukin-1 and tumor necrosis factor α in rheumatoid arthritis. *Arthritis and Rheumatology* 38(2):151–160.

Arroyo, C., Eslami, S., Brunton, N.P., Arimi, J.M., Noci, F. and Lyng, J.G. 2015. An assessment of the impact of pulsed electric fields processing factors on oxidation, color, texture, and sensory attributes of turkey breast meat. *Poultry Science* 94(5):1088–1095.

Barrow, C., Wang, B., Adhikari, B. and Liu, H. 2013. Spray drying and encapsulation of omega-3 oils. In: Jacobsen C, Nielsen NS, Horn AF, Sorensen M, editors. *Food Enrichment with Omega-3 Fatty Acids*. Woodhead Publishing Ltd.: Oxford, UK. pp 194–225.

Becalski, A. and Seaman, S. 2005. Furan precursor in food: A model study and development of a simple headspace method for determination of furan. *Journal of AOAC International* 88(1):102–106.

Berlitz, H.-D., Grosche, W. and Schieberle, P. 2009. Lipids. In: Berlitz H-D, Grosch W, Schieberle P, editors. *Food Chemistry*. Springer: Berlin, Germany. pp 158–247.

Byrne, D.V. 2000. Sensory characterisation studies on warmed-over flavour in meat. PhD Thesis. Department of Dairy and Food Science, Royal Veterinary and Agricultural University: Copenhagen, Denmark.

Calder, P.C. 2013. Nutritional benefits of omega-3 fatty acids. In: Jacobsen NS, Nielsen AF, Horn AD, Sorensen M, editors. *Food Enrichment with Omega-3 Fatty Acids.* Woodhead Publishing Ltd: Cambridge, UK.

de Lorgeril, M., Salen, P., Laporte, F. and de Leiris, J. 2001. Alpha-linolenic acid in the prevention and treatment of coronary heart disease. *European Heart Journal Supplement D* 3:D26–D32.

Fan, X. 2015. Furan formation from fatty acids as a result of storage, gamma irradiation, UV-C and heat treatments. *Food Chemistry* 175:439–444.

Flaskerud, K., Bukowski, M., Golovko, M., Johnson, L., Brose, S., Ali, A., Cleveland, B., Picklo, M. and Raatz, S. 2017. Effects of cooking techniques on fatty acid and oxylipin content of farmed rainbow trout (*Oncorhynchus mykiss*). *Food Science and Nutrition* 5(6):1195–1204.

Fournier, V., Destaillats, F., Juanéda, P., Dionisi, F., Lambelet, P., Sébédio, J.L. and Berdeaux, O. 2006. Thermal degradation of long-chain polyunsaturated fatty acids during deodorization of fish oil. *European Journal of Lipid Science and Technology* 108(1):33–42.

Gharsallaoui, A., Roudaut, G., Chambin, O., Voilley, A. and Saurel, R. 2007. Applications of spray-drying in microencapsulation of food ingredients: An overview. *Food Research International* 40(9):1107–1121.

Ghnimi, S., Budilarto, E. and Kamal-Eldin, A. 2017. The new paradigm for lipid oxidation and insights to microencapsulation of Omega-3 fatty acids. *Comprehensive Reviews in Food Science and Food Safety* 16(6):1206–1219.

Gladyshev, M.I., Sushchik, N.N., Gubanenko, G.A., Demirchieva, S.M. and Kalachova, G.S. 2006. Effect of way of cooking on content of essential polyunsaturated fatty acids in muscle tissue of humpback salmon (*Oncorhynchus gorbuscha*). *Food Chemistry* 96(3):446–451.

Gomez-Mascaraque, L. and Lopez-Rubio, A. 2016. Protein-based emulsion electrosprayed micro- and submicroparticles for the encapsulation and stabilization of thermosensitive hydrophobic bioactives. *Journal of Colloid and Interface Science* 465:259–270.

Hadaruga, D.I., Unlusayin, M., Gruia, A.T., Birau, C., Rusu, G. and Hadaruga, N.G. 2016. Thermal and oxidative stability of Atlantic salmon oil (*Salmo salar* L.) and complexation with β-cyclodextrin. *Beilstein Journal of Organic Chemistry* 12:179–191.

He, Z., Zeng, W., Zhu, X., Zhao, H., Lu, Y. and Lu, Z. 2017. Influence of surfactants on physical and oxidative stability of microemulsions with docosahexaenoic acid. *Colloids and Surfaces, Part B: Biointerfaces* 151:232–239.

Hrncirik, K. 2010. Stability of fat-soluble vitamins and PUFA in simulated shallow-frying. *Lipid Technology* 22(5):107–109.

Huang, J. and Sathivel, S. 2008. Thermal and rheological properties and the effects of temperature on the viscosity and oxidation rate of unpurified salmon oil. *Journal of Food Engineering* 89(2):105–111.

Jacobsen, C. 2015. Some strategies for the stabilization of long chain n-3 PUFA-enriched foods: A review. *European Journal of Lipid Science and Technology* 117(11):1853–1866.

Kajla, P., Sharma, A. and Sood, D.R. 2015. Flaxseed—A potential functional food source. *Journal of Food Science and Technology* 52(4):1857–1871.

Kavousi, H.R., Fathi, M. and Goli, S.A.H. 2017. Stability enhancement of fish oil by its encapsulation using a novel hydrogel of cress seed mucilage/chitosan. *International Journal of Food Properties* 20(2):1890–1900.

Kremer, J.M. 2000. n-3 fatty acid supplements in rheumatoid arthritis. *The American Journal of Clinical Nutrition* 71:349S–351S.

Larsson, K., Harrysson, H., Havenaar, R., Alminger, M. and Undeland, I. 2016. Formation of malondialdehyde (MDA), 4-hydroxy-2-hexenal (HHE) and 4-hydroxy-2-nonenal (HNE) in fish and fish oil during dynamic gastrointestinal in vitro digestion. *Food Function* 7(2):1176–1187.

Let, M.B., Jacobsen, C. and Meyer, A.S. 2005. Sensory stability and oxidation of fish oil enriched milk is affected by milk storage temperature and oil quality. *International Dairy Journal* 15(2):173–182.

Leung, K.S., Galano, J.M., Durand, T. and Lee, J.C.Y. 2018. Profiling of omega-polyunsaturated fatty acids and their oxidized products in salmon after different cooking methods. *Antioxidants* 96(7):1–13.

Liu, F., Zhu, Z., Ma, C., Luo, X., Bai, L., Decker, E.A., Gao, Y. and McClements, D.J. 2016. Fabrication of concentrated fish oil emulsions using dual-channel microfluidization: Impact of droplet concentration on physical properties and lipid oxidation. *Journal of Agricultural and Food Chemistry* 64(50):9532–9541.

Ludger, B. 2014. Fatty acid alterations in oils and fats during heating and frying. *European Journal of Lipid Science and Technology* 116S:707–715.

Mallia, S., Escher, F., Dubois, S., Schieberle, P. and Schlichtherle-Cerny, H. 2009. Characterization and quantification of odor-active compounds in unsaturated fatty acid/conjugated linoleic acid (UFA/CLA)-enriched butter and in conventional butter during storage and induced oxidation. *Journal of Agricultural and Food Chemistry* 57(16):7464–7672.

McDonnell, C., Allen, K., Chardonnereau, P., Arimi, F.S., J.M. and Lyng, J.G. 2014. The use of pulsed electric fields for accelerating the salting of pork. *Food Science and Technology* 59:1054– 1060.

Pankaj, S.K., Bueno-Ferrer, C., and Misra, N. 2014. Applications of cold plasma technology in food packaging. *Trends in Food Science and Technology* 35(1):5–17.

Punia, S., Sandhu, K.S., Sandhu, Siroha, A.K. and Dhull, S.B. 2019. Omega-3 metabolism, absorption, bioavailability and health benefits-A review. *Pharma Nutrition*. doi:10.1016/j.phanu.2019.100162.

Sharoba, A.M. and Ramadan, M.F. 2012. Impact of frying on fatty acid profile and rheological behaviour of some vegetable oils. *Journal of Food Processing and Technology* 2(3(7)):1–9.

Shaw, L.A., McClements, D.J. and Decker, E.A. 2007. Spray-dried multilayered emulsions as a delivery method for omega-3 fatty acids into food systems. *Journal of Agricultural and Food Chemistry* 55(8):3112–3119.

Sioen, I., Haak, L., Raes, K., Hermans, C., De Henauw, S., De Smet, S. and Van Camp, J. 2006. Effects of pan-frying in margarine and olive oil on the fatty acid composition of cod and salmon. *Food Chemistry* 98(4):609–617.

Stephen, N.M., Jeya Shakila, R., Jeyasekaran, G. and Sukumar, D. 2010. Effect of different types of heat processing on chemical changes in tuna. *Journal of Food Science and Technology* 47(2):174–181.

Velasco, J., Marmesat, S., Dobarganes, C. and Marquez-Ruiz, G. 2006. Heterogeneous aspects of lipid oxidation in dried microencapsulated oils. *Journal of Agricultural and Food Chemistry* 54(5):1722–1729.

Walker, R., Decker, E.A. and McClements, D.J. 2015. Development of food-grade nanoemulsions and emulsions for delivery of omega-3 fatty acids: Opportunities and obstacles in the food industry. *Food and Function* 6(1):42–55.

Yepez, X.V. and Keener, K.M. 2016. High-voltage atmospheric cold plasma (HVACP) hydrogenation of soybean oil without trans-fatty acids. *Innovative Food Science and Emerging Technologies* 38:169–174.

Zhang, Q., Saleh, A.S.M., Chen, J. and Shen, Q. 2012. Chemical alterations taken place during deep fat frying based on certain reaction products: A review. *Chemistry and Physics of Lipids* 165(6):662–681.

7

Effect of Storage on Essential Fatty Acids

Parveen Kumari

Contents

7.1 INTRODUCTION

Polyunsaturated fatty acids (PUFAs) collectively include unsaturated fatty acids containing two or more double bonds (Calder, 2008). PUFAs are grouped into two categories, omega-3 and omega-6, on the basis of the last double bond position relative to the methyl end terminal of the fatty acid (Wall et al., 2010). Alpha-linolenic acid and linoleic acid are referred to as essential fatty acids (EFAs) because the human body cannot synthesize these fatty acids. The essential fatty acids are the precursors of long-chain highly unsaturated fatty acids like C20 and C22. Recently, sources of omega-3 have been a major focus of scientific research due to their various benefits for the metabolism and health (Punia et al., 2019). The body requirement of mammals regarding the n-6 series is more than that for the n-3 series. However, intense ingestion of omega-6 compared to omega-3 might lead to insufficiency of docosahexaenoic acid and eicosapentaenoic acid. Both fatty acids are well known for their valuable health effects, particularly to avoid various pathologies such as cardiovascular diseases (Lovegrove and Griffin, 2013; Delgado-Lista et al., 2012). Recently, conjugated dienoic fatty acids have been attracting the attention of researchers and nutritionists, being a potent source of anti-atherogenic and anti-cancer properties. PUFAs are more susceptible to oxidative deterioration or auto-oxidation, generally favored by thermal processing and storage conditions (Mazalli and Bragagnolo, 2009). The stability of EFA during storage is condition dependent and generally influenced by various factors like the composition of lipids and the types of fatty acids in products, the level of pro-oxidants, the degree of unsaturation, the position of double bonds, enzymes, iron (Fe^{+2}) concentration, the temperature of storage, pH, oxygen level and ionic strength (Skibsted, Mikkelsen and Bertelsen, 1998). Lipid oxidation is the main chemical reaction in stored food that occurs due to the autocatalytic breakdown of UFAs, generally initiated by the interaction of hydrogen atoms in fatty acid chain and singlet oxygen (Steger and Muhlebach, 1997). Hydroxyl radicals further react with fat and produce peroxyl radicals, which leads to hyper-peroxide production (Slater, 1984). Peroxide breakdown might result in the production of aldehydes and other polymerization products (Liang, 1999). The oxidation of lipids results in the production of off-odor and off-flavor. The assessment of fatty acid profiles and understanding the storage stability are very complex; the main hindrance is the dependence of analytical techniques on the nature of samples and fatty acid types.

7.2 STORAGE EFFECT ON EFA FROM MARINE SOURCES

PUFAs with numerous unsaturated bonds are more susceptible to oxidation compared to MUFAs and SFAs (Yi-Chen et al., 2008). Oxidation is categorized into two groups: enzymatic and non-enzymatic oxidation (auto-oxidation and photo-oxidation). Lipooxygenases, cyclooxygenases and cytochrome P 450 are the main enzymes involved in the enzymatic oxidation of PUFAs. Auto-oxidation and photo-oxidation are mediated by free radicals and ultraviolet or singlet oxygen, respectively (Gruber et al., 2015) (Figure 7.1).

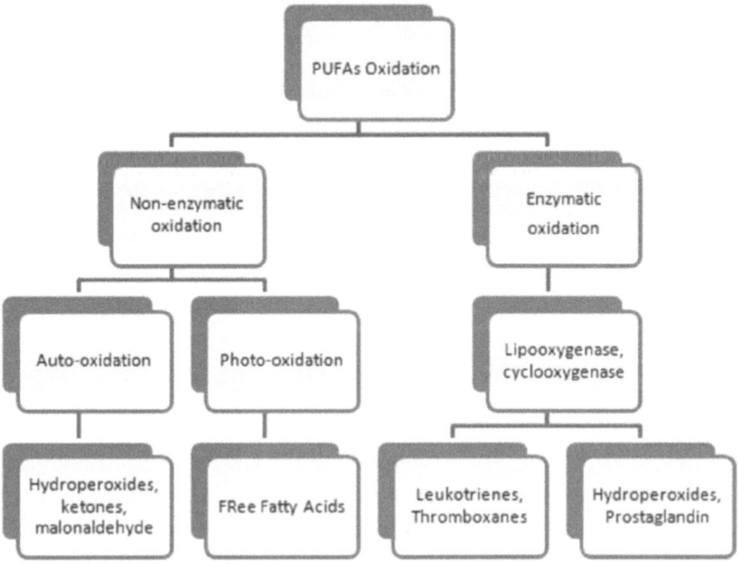

Figure 7.1 Oxidation of polyunsaturated fatty acids

Hydroperoxides, primary oxidation products of auto-oxidation, further oxidized to ketones and malonaldehyde. Hydroxyl alkenals like 4-hydroxy-2-hexenal and 4-hydroxy-2-nonenal are formed by oxidation of n-3 and n-6, respectively. The photo-oxidation of PUFAs includes photosensitization and photochemical oxidation. Photosensitization oxidation is initiated in the presence of photosensitizers like chlorophyll, riboflavin, hemeprotein and synthetic dyes along with visible light, while photochemical oxidation requires ultra irradiation exposure (Tao, 2015). Vegetable oils containing chlorophyll pigment are more prone to oxidation during storage. PUFAs liberated by the activity of enzyme phospholipases A2 on phospholipids can be further catalyzed by lipooxygenases and cyclooxygenases to produce hydroperoxides and prostaglandin H2, respectively. Recent studies on PUFA oxidation reported that lipooxygenases catalyzed products as potent fat mediators. E-series and D-series resolvins are new metabolites produced by enzymatic oxygenation of EPA and DHA, respectively (Ward, 2015). Lipooxygenases and cyclooxygenases can also act on omega-3 and omega-6 PUFAs to generate series of prostaglandins, leukotrienes and thromboxanes (Schmitz and Ecker, 2008).

$$\text{PUFAs} \xrightarrow{\text{Oxidation}} \text{Fatty acid peroxides} + \text{Malonaldehydes} \xrightarrow[\text{and amino acids}]{\text{Nucleosides, proteins}} \text{TBARS}$$

$$\text{PUFAs} \xrightarrow{\text{Hydrolysis}} \text{Free fatty acids} \xrightarrow[\text{and amino acids}]{\text{Nucleosides, proteins}} \text{TBARS}$$

Hydrolysis and oxidation can be catalyzed by various types of enzymes, such as microbial, digestive or intracellular (Richards and Hultin, 2002).

Omega-3 long-chain PUFAs (EPA and DHA) contained more bisallylic carbons than short-chain omega-3 PUFA (ALA) and n-6 (arachidonic acid). Carbon between two double-bonded carbons is known as bisallylic carbon and requires low activation energy for hydrogen liberation and fatty acid production. Different initiators give rise to lipid radical formation that further produces peroxides and more radicals from unoxidized polyunsaturated fatty acids. Numerous complex peroxide molecules are formed depending on the position of carbons and after *cis-trans* isomerization and a swing of double bond. Conjugated dienes and trienes produced after these reactions differ in shape and polarity from base fatty acids (Shahidi and Zhong, 2010).

Prostaglandin-like F3-isoprostanes and F4-neuroprostanes are the peroxides produced by EPA and DHA oxidation, respectively (Barden et al., 2011). Peroxides, being highly unstable, undergo secondary oxidation to form aldehydes like malondialdehyde and 4-hydroxyhexenal (Esterbauer, 1993). As oil oxidized with storage time, the concentration of lipid peroxides increased exponentially, and the concentration was later reduced with an increase in the production of harmful secondary metabolites. Omega-3 long-chain PUFAs stored under light are more prone to oxidation in the presence of proteins, heavy metals and their conjugates (Shahidi and Zhong, 2010). Light is a major factor responsible for spoilage of fats and oils. As fat itself does not absorb the visible light spectrum, it might be induced light absorbed by impurities in oils and fats (Yenge et al., 2017).

Storage under frozen conditions is considered to be effective to enhance the shelf life of fish with good nutritional value, as it inhibits microbial growth and reduces various biochemical processes, viz. hydrolysis and lipid oxidation, responsible for the deterioration of seafood quality. Lipid oxidation is an important reaction in fat-rich foods that determines functional, nutritional and organoleptic properties along with their storage stability. Hydroperoxides accumulate as primary oxidation products of oxidation that further decomposed to secondary metabolites like FFA, alcohols, ketones and aldehydes, resulting in rancid flavor (Farhoosh et al., 2009). Linolenic acids are comparatively more prone to auto-oxidation than linoleic and oleic acids due to having more methylene groups showing easy cleavage of hydrogen atoms (Shahidi and Zhong, 2010).

Being an excellent source of omega-3 PUFAs, mackerel is considered very nutritive. Docosahexaenoic acid and eicosapentaenoic acid are the dominating PUFAs, followed by stearidonic acid, alpha-linoleic acid and linoleic acid. Mackerel from Icelandic waters was stored at different frozen temperatures (–18°C and –25°C). The PUFA value was higher in fish caught in August than in fish caught in September. The storage stability of lipid was estimated by hydrogen peroxide, thiobarbituric acid reactive substances (TBARS), FFA value and alteration in FA composition. Romotowska et al. (2016) observed no significant change in DHA level while EPA showed a slight decrease over 9 months of storage at both the storage conditions. Primary oxidation products, i.e., hydrogen peroxide value (PV), increased significantly with time of storage from 6 to 12 months. Similarly, TBARS (secondary oxidation product content) increased with storage period. Free fatty acid resulting from lipid oxidation showed similar trends of increment like PV and TBARS at both conditions: samples at –25°C were more acceptable than those at –18°C. Romotowska et al. (2016) demonstrated higher stability of PUFAs at –25°C compared to –18°C during long-term storage along with better retention of n-3 fatty acids in frozen mackerel and its products.

Nazemroaya et al. (2009) analyzed the effect of freezing (–30°C) on FA composition, lipid oxidation and enzymatic breakdown of lipids on fish fillets caught from the Persian Gulf. Marine lipids are excellent sources of PUFAs, specially DHA and EPA. White cheek shark and narrow-barred Spanish mackerel were tested at regular intervals of 1 month during storage for 6 months at –18°C. In both marine species, the PUFA + MUFA levels were higher than SFAs, whereas a decline was reported more in UFAs compared to SFAs during storage under frozen conditions. Both linoleic acid and arachidonic acid were major FAs in total n-6 PUFAs in both shark and mackerel. The EPA and DHA were the main fatty acids among all the n-3 acids. PUFA levels in mackerel and shark decreased 40.1% and 23.94% (n-3 (48% and 42.83%), x3/x6 (41.36% and 50%), EPA + DHA / C16 (55.55% and 46.66%)), respectively. Peroxide, TBA (thiobarbituric acid) and FFA values showed significant increases correlating with length of storage. Malonaldehyde, a by-product of lipid oxidation, is reported in the form of the TBA number. A significant increase was noted in TBA from 0.038 to

0.068 mg (shark) and from 0.043 to 0.104 mg (mackerel) of malonaldehyde/kg of flesh during storage. TBA values might not present the exact rate of lipid oxidation, as malonaldehyde might interact with other compounds like proteins, nucleosides, amino acids of phospholipids, amines or lipid oxidation products and interaction might differ widely according to fish species (Namulema et al., 1999).

Lipid hydrolysis favored production of FFA in both fishes stored under frozen conditions. Peroxide values increased from 0.24 to 4.43 and from 2.32 to 15.41 meq of O_2/kg lipid in shark and mackerel, respectively, during storage. During the 1st month PV was constant, while a sharp decline was reported in the 6th month. However, storage under frozen conditions reduces microbial deterioration, and muscle proteins undergo numerous modifications, including insolubility and aggregate formation, that further altered their structure and functional characteristics. The oxidation of lipids continued during frozen storage, resulting in qualitative and nutritional changes, and varied according to the levels of unsaturated fatty acids, which are highly susceptible to peroxidation and rapid deterioration (Badii and Howell, 2002). The study suggested shorter frozen storage periods for better retention of nutrition and sensory qualities in frozen fish.

Chávez-Mendoza et al. (2014) stored 54 rainbow trout fillets at –15°C for 45 and 90 days and reported the storage effect on the fatty acid profile. Fresh fillets have good omega-3 content, including DHA (39%), EPA (5.64%), DPA (1.77%) and alpha-linolenic acid (1.09%). PUFA, omega-6 and omega-3 content decreased by 25.3%, 12.4% and 32.55%, respectively, while increases were observed in SFA (32.63%) and MUFA (9.25%) contents during the first 45 days of storage. At the end of 90 d of storage, DHA, DPA, EPA and alpha-linolenic acid decreased by 38.71%, 35.79%, 26.77% and 7.27%, respectively. Overall, frozen storage reduced the nutritional profile of fillets due to declines in PUFA level and n-3/n-6 ratio and an increase in SFA. Correlation results concluded that the concentration of fatty acid and the quality index were both frozen condition dependent. PUFAs were negatively correlated with frozen storage time.

Memon et al. (2010) analyzed FFA, peroxide value, acid value, iodine value, sponification value, polyene index (PI) and fatty acid profile of freshwater fish oil Jarko (*Wallago attu*) stored under a refrigeration condition for 120 days. Among PUFAs, DHA was the main one, followed by EPA. A decline in PUFA (33.70%) and MUFA (24.89%) content was reported during storage while SFAs increased by 26.82%. PUFA content was measured by polyene index during storage. PI value was reduced due to lipid oxidation, but the value remained constant after the 90th day. PI might be an important tool to analyze the oxidative stability of fish oils, as they contain two main LC-PUFAs (essential fatty acids) and thus their level can be estimated with reasonable accuracy. Among PUFAs, EPA (43.37%) was decreased more compared to DHA (16.43%) during 120 days of storage and that might be due to their high susceptibility to oxidation. Fish oil on the basic of PV and FFA values remained acceptable for 60 days but eventually crossed recommended values after 60 days of storage.

7.3 STORAGE EFFECT ON EFA IN EGGS

Douny et al. (2015) investigated the effect of storage (4°C and 20°C for 6 weeks) on raw and cooked Columbus eggs. According to European Parliament and Council Directive No. 1924/2006, Columbus eggs can be documented as "high omega-3 fatty acid" products, as they contain >0.6 g LNA/100 g (based on 7.5% fat containing 12% LNA, egg has approximately 0.9 g of LNA/100 g). Columbus eggs have higher n-3 LNA fatty acid—about 12.7% compared to normal egg (1.1%)—and therefore, PUFA content was high (30.2%) compared to standard egg (20.2%). No significant change was reported in the ARA, DHA and LNA contents of stored egg at +4°C and +20°C for 6 weeks compared to control. Studies by other researchers reported no significant change in fatty acid composition of n-3 enriched eggs during storage at room temperature for 28 days (Meluzzi et al., 2000). Egg cooked in custard showed a slight decline in omega-3 fatty acids (11.1%) compared to raw egg (15.3%).

Ren et al. (2013) examined the effect of cooking and retail storage conditions on omega-3 PUFA-enriched eggs supplemented with antioxidants, viz. vitamin E and organic selenium. Total fat content showed no significant variations while n-3 FAs were reduced significantly. Cooking techniques significantly affected lipid oxidation in eggs. Malondialdehyde content and cholesterol oxidation products (COPs) were higher in fried eggs followed by boiled and raw eggs. Antioxidant fortification reduced MDA formation by 40% and COPs by 12% in fried eggs. The linolenic acid and linoleic acid content of omega-3-enriched eggs were not affected during a 28-day storage period, while long-chain PUFAs such as DPA, DHA, EPA and arachidonic acid were reduced by 12%, 8.74%, 6.74% and 4.72%, respectively. Omega-3 fatty acids were reduced more compared to omega-6; therefore, the ratio of n-6/n-3 showed an increment from 1.93 to 1.95 after a storage period of 28 days. In fresh raw egg, the MDA content varied from 0.32 to 0.42 mg/kg sample.

Mazalli and Bragagnolo (2009) reported the effect of storage (5°C and 25°C for 45 days) on FA composition of n-3 enriched eggs (raw, boiled and fried). PUFA (18:3, 20:5 and 22:6) and MUFA content decreased in thermal treatments as well as during storage under both conditions. Peroxides and free acids generated due to lipid oxidation accelerated the formation of cholesterol oxides during storage in eggs.

7.4 EFFECT OF STORAGE ON EFA FROM VEGETABLE OILS

Soybean fat has an exceptionally high content of PUFAs, viz. linolenic acid and linoleic acid. The beany taste of soyabean product is a common issue associated with the oxidation of UFAs by lipoxygenase to produce volatile compounds (Rhee et al., 1989). Prabakaran et al. (2018) studied the storage effect on stability and quality of soya flour stored at refrigeration condition (4°C), room temperature (20°C), and elevated temperature (45°C). Results showed that major UFAs of raw soyabean flour like linolenic, linoleic and oleic acids reduced by 97.7%, 94.4% and 30%, respectively, stored at high temperature for 48 hours. Values were enhanced later due to hydrolysis in flour. The most impact on fatty acid stability was from the storage period, followed by the storage temperature and the packaging material. Hexanal, 2,4-heptadienal and nonadienal are formed as secondary metabolites of fatty acid degradation and act as indicators of oxidation in food products rich in fat (Fu et al., 2009).

El Haouhay et al. (2018) studied the effect of storage on the quality of virgin olive oil extracted in traditional mills in Morocco. Samples packed in dark glass and PET bottles were analyzed after 3 and 6 months. Two main PUFAs, linoleic and linolenic acids, present in olive oil showed reduction during the storage period. The PUFA content of sunflower oil stored at 25°C for 3 months decreased by 12.4%, while MUFA and SAF were increased by 30.8% and 5.2%, respectively. The peroxide value of oil increased by 72.1 meq $O_2 \cdot kg^{-1}$ during the storage period. Sunflower oil fortified with green coffee extract (3 g/100 g) had better retention of PUFAs and PV was increased by 37.9 meq $O_2 \cdot kg^{-1}$. Therefore, green coffee extract might serve as an antioxidant in sunflower oil during the storage period (Budryn, et al., 2014).

Zupo et al. (2015) studied the storage effect (12 months at –20°C) on a fat diet formulated for post-larvae of the crustacean decapods *Hippolyte inermis*. Stearic acid, oleic acid, linoleic acid and EPA showed stability during storage, while significant reduction was reported in α-linolenic acid, γ-linolenic acid and arachidonic acid.

7.5 EFFECT OF STORAGE ON EFA FROM HUMAN MILK AND OTHER FORMULATIONS

Rodriguez-Alcala et al. (2007) reported on the effect of storage under ambient conditions (25 ± 3°C) for 4 years on the fatty acid profile of powdered infant formula. A potential decline was noticed in linolenic acid content during the first year of storage, while linoleic acid showed significant reduction in the fourth month of storage. Reduction was mainly attributed to peroxide formation during the storage period. No oxidative products or *trans* fatty acids were noticed in the stored formulation, and the loss in tocopherol content was not significant. Therefore, the lipid profile of the infant formulation was not altered greatly during long-term storage.

Human milk stored at 4°C for 6 days and at –20°C and –80°C for 6 months was analyzed for LA, ALA, ARA and DHA along with vitamin C and riboflavin content. No alteration was reported in four fatty acid levels in the sample stored at 4°C for 6 days. Abramovich et al. (2013) recommended storage of human milk at either –20°C or –80°C for better preservation of these four fatty acids and two vitamins. The storage life might be extended with Holder pasteurization. In milk, the fat globule membrane has a protective role for fatty acids, as these are incorporated in globules (Abramovich et al., 2013).

7.6 CONCLUSIONS

Consumption of a diet enriched with LC omega-3 and omega-6 PUFAs, especially DHA and EPA, is believed to be linked with reduced risks of cardiovascular and Alzheimer's disease. The stability of essential fatty acids depends on source, types and composition of lipids, the physical and colloidal stage of fat, storage conditions and the presence of antioxidants. Lipid oxidation affects the quality of fish and other oil-based food by generating rancid off-flavor and by reducing nutritional quality. Storage under frozen conditions, especially at –18°C or –25°C, retained fatty acid profiles for shorter durations. Several studies suggesting the application of antioxidants during the storage of PUFAs reduced lipid oxidation.

Antioxidants, modified atmospheric storage packaging and encapsulation can be applied to reduce or delay lipid oxidation. Microencapsulation involves incorporation of some sensitive ingredients like omega-3 PUFAs into food without altering the sensory qualities of food. Encapsulated freeze-dried oils can be stored for longer durations with a better nutritional profile. Further storage studies on foods rich in PUFAs are greatly needed for better nutrient retention for longer periods. The application of encapsulation and microencapsulation technology along with freezing might be helpful for improved PUFA storage.

REFERENCES

Abramovich, M., Friel, J. K. and Hossain, Z. 2013. Polyunsaturated fatty acids, riboflavin and vitamin C: Effect of different storage conditions of human milk. *Vitamins and Minerals*. 2(1): 110. doi:10.4172/vms.1000110.

Badii, F. and Howell, N. K. 2002. Changes in the texture and structure of cod and haddock fillets during frozen storage. *Food Hydrocolloids*. 16(4): 313–319.

Barden, A., Mas, E. and, Henry, P. 2011. The effects of oxidation products of arachidonic acid and n3 fatty acids on vascular and platelet function. *Free Radical Research*. 45(4): 469 476.

Budryn, G., Nebesny, E., Żyżelewicz, D. and Oracz, J. 2014. Properties of model systems of sunflower oil and green coffee extract after heat treatment and storage. *LWT - Food Science and Technology*. 59(1): 467–478. doi:10.1016/j.lwt.2014.05.048.

Calder, P. C. 2008. Poly unsaturated fatty acids, inflammatory processes and inflammatory bowel diseases. *Molecular Nutrition and Food Research*. 52(8): 885–897.

Chávez-Mendoza, C., García-Macías, J. A., Alarcón-Rojo, A. D., Ortega-Gutiérrez, J. Á., Holguín-Licón, C. and Corral-Flores, G. 2014. Comparison of fatty acid content of fresh and frozen fillets of rainbow trout (*Oncorhynchus mykiss*) *Walbaum*. *Brazilian Archives of Biology and Technology*. 57(1): 103–109.

Delgado-Lista, J., Perez-Martinez, P., Lopez-Miranda, J. and Perez-Jimenez, F. 2012. Long chain omega-3 fatty acids and cardiovascular disease: A systematic review. *The British Journal of Nutrition*. 107: S201–213.

Douny, C., El Khoury, R., Delmelle, J., Brose, F., Degand, G., Moula, N., Farnir, F., Clinquart, A., Maghuin-Rogister, G. and Scippo, M. L. 2015. Effect of storage and cooking on the fatty acid profile of omega-3 enriched eggs and pork meat marketed in Belgium. *Food Science and Nutrition*. 3(2): 140–152.

El-Haouhay, N., Samaniego-Sánchez, C., Asehraou, A., Giménez-Martínez, R. J., Villalón-Mir, M. and Serrana, H. L.-G. De la 2018. Effects of olive storage and packaging on microbial and fatty acids profiles of olive oil produced in traditional mills in Morocco. *Journal of Materials and Environmental Science*. 9(3): 854–863.

Esterbauer, H. 1993. Cytotoxicity and genotoxicity of lipid-oxidation products. *The American Journal of Clinical Nutrition*. 57(5): 779S–785S.

European Parliament and Council Directive no 1924/2006. 2006. European Parliament and Council Directive No 1924/ 2006 of 20 December 2006 on nutrition and health claims made on foods. *Official Journal of European Communities*. L 404: 9–25.

Farhoosh, R., Einafshar, S. and Sharayei, P. 2009. The effect of commercial refining steps on the rancidity measures of soybean and canola oils. *Food Chemistry*. 115(3): 933–938.

Fu, X., Xu, S. and Wang, Z. 2009. Kinetics of lipid oxidation and off-odor formation in silver carp mince: The effect of lipoxygenase and hemoglobin. *Food Research International*. 42(1): 85–90.

Gruber, F., Ornelas, C. M. and Karner, S. 2015. Nrf2 deficiency causes lipid oxidation, inflammation, and matrix-protease expression in DHA-supplemented and UVA-irradiated skin fibroblasts. *Free Radical Biology and Medicine*. 88(B): 439–451.

Liang, J. H. 1999. Fluorescence due to interactions of oxidizing soybean oil and soy proteins. *Food Chemistry*. 66(1): 103–108.

Lovegrove, J. A. and Griffin, B. A. 2013. The acute and longterm effects of dietary fatty acids on vascular function in health and disease. *Current Opinion in Clinical Nutrition and Metabolic Care*. 16(2): 162–167.

Mazalli, M. R. and Bragagnolo, N. 2009. Increase of cholesterol oxidation and decrease of PUFA as a result of thermal processing and storage in eggs enriched with n-3 fatty acids. *Journal of Agricultural and Food Chemistry*. 57(11): 5028–5034.

Meluzzi, A., Sirri, F., Manfreda, G., Tallarico, N. and Franchini, A. 2000. Effects of dietary vitamin E on the quality of table eggs enriched with n-3 long-chain fatty acids. *Poultry Science*. 79(4): 539–545.

Memon, N. N., Talpur, F. N., Sherazi, S. T. H. and Bhanger, M. I. 2010. Impact of refrigerated storage on quality of oil from freshwater Jarko (*Wallago attu*) fish. *Pakistan Journal of Analytical and Environmental Chemistry*. 11(2): 37–43.

Namulema, A., Muyonga, J. H. and Kaaya, N. A. 1999. Quality deterioration in frozen Nile perch (*Lates niloticus*) stored at –13 and –27°C. *Food Research International*. 32(2): 151–156.

Nazemroaya, S., Sahari, M. A. and Rezaei, M. 2009. Effect of frozen storage on fatty acid composition and changes in lipid content of *Scomberomorus commersoni* and *Carcharhinus dussumieri*. *Journal of Applied Ichthyology*. 25(1): 91–95.

Prabakaran, M., Lee, K.-J., An, Y., Kwon, C., Kim, S., Yang, Y., Ahmad, A., Kim, S.-H. and Chung, M. 2018. Changes in soybean (glycine max l.) flour fatty-acid content based on storage temperature and duration. *Molecules*. 23(10): 2713. doi:10.3390/molecules23102713.

Ren, Y., Perez, T. I., Zuidhof, M. J., Renema, R. A. and Wu, J. 2013. Oxidative stability of Omega-3 polyunsaturated fatty acids enriched eggs. *Journal of Agricultural and Food Chemistry*. 61(47): 11595–11602.

Rhee, C. O., Kim, D. Y., Jung, J. H., Kim, K., Park, K. H. and Chung, H. J. 1989. Effects of cooking conditions on the texture of cooked soybeans. *Journal of Korean Agricultural Chemical Society*. 32(3): 216–221.

Richards, P. M. and Hultin, O. H. 2002. Contributions of blood and blood components to lipid oxidation in fish muscle. *Journal of Agricultural and Food Chemistry*. 50(3): 555–564.
Rodriguez-Alcala, L. M., Garcia-Martinez, M. C., Cachon, F., Marmesat, S., Alonso, L., Marquez-Ruiz, G. and Fontecha, J. 2007. Changes in the lipid composition of powdered Infant Formulas during long-term storage. *Journal of Agricultural and Food Chemistry*. 55(16): 6533–6538.
Romotowska, P. E., Karlsdottir, M. G., Gudjonsdottir, M., Kristinsson, H. G. and Arason, S. 2016. Influence of feeding state and frozen storage temperature on the lipid stability of Atlantic mackerel (Scomber scombrus). *International Journal of Food Science and Technology*. doi:10.1111/ijfs.13146.
Schmitz, G. and Ecker, J. 2008. The opposing effects of n-3 and n-6 fatty acids. *Progress in Lipid Research*. 47(2): 147–155.
Shahidi, F. and Zhong, Y. 2010. Lipid oxidation and improving the oxidative stability. *Chemical Society Reviews*. 39(11): 4067–4079.
Skibsted, L. H., Mikkelsen, A. and Bertelsen, G. 1998. Lipid-derived off-flavours in meat. In: F. Shahidi (Ed.), *Flavor of Meat, Meat Products and Sea Foods*. (2nd ed., pp. 217–256). London: Blackie Academic.
Slater, T. F. 1984. Overview of methods used for detecting lipid peroxidation. *Methods in Enzymology*. 105: 273–292.
Punia, S., Sandhu, K. S., Siroha, A. K. and Dhull, S. B. 2019. Omega 3- Metabolism, absorption, bioavailability and health benefits: A review. *Pharma Nutrition*. doi:10.1016/j.phanu.2019.100162.
Steger, P. J. K. and Mühlebach, S. F. 1997. In vitro oxidation of IV lipid emulsions in different all-in-one admixture bags assessed by an iodometric assay and gas-liquid chromatography. *Nutrition*. 13(2): 133–140.
Tao, L. 2015. Oxidation of polyunsaturated fatty acids and its impact on food quality and human health. *Advances in Food Technology and Nutritional Sciences-Open Journal*. 1(6): 135–142.
Wall, R., Ross, R. P., Fitzgerald, G. F. and Stanton, C. 2010. Fatty acids from fish: The anti-inflammatory potential of long chain omega-3 fatty acids. *Nutrition Reviews*. 68(5): 280–289.
Ward, P. A. 2015. Resolvins on the way to resolution. *The Journal of Experimental Medicine*. 212(8): 1142.
Yenge, G. B., More, H. G., Thorat, S. S., Naik, R. M. and Nimbalkar, C. A. 2017. Storage study of garden cress (*Lepidium sativum* L.) seed oil. *Journal of Pharmacognosy and Phytochemistry*. 6(3): 09–13.
Yi-Chen, Ch., Nguyen, J., Semmens, K., Beamer, S. and Jaczynski, J. 2008. Chemical changes in omega-3 enhanced farmed rainbow trout (*Oncorhynchus mykiss*) fillets during abusive-temperature storage. *Food Control*. 19(6): 599–608.
Zupo, V., Maibam, C., Fink, P. and Elert, E. 2015. Effect of storage on the fatty acid content of foods for post-larvae of the crustacean decapods *Hippolyte inermis*. *Invertebrate Reproduction and Development*. 59(2): 1–10.

8

Essential Fatty Acid Bioavailability

A Dietary Perspective

Mohammad Kashif Kidwai, Ajay Singh, Tanu Malik,
Sanju Bala Dhull and Sneh Punia

Contents

8.1 INTRODUCTION

Functional foods, in terms of essential fatty acids (EFAs), include the polyunsaturated fatty acid (PUFA) class of fats; PUFAs are essential for the body to be fit and fine. The introductory classification of fatty acids is based on their chemical structure (Nettleton, 1995) and the fact that they are required by the human body, a requirement that can only be satisfied through dietary means (Punia et al., 2019). Their essentiality is defined by the way that the human body is incompetent to synthesize them, as the human metabolism is not efficient enough to incorporate double bonds in the requisite locations of hydrocarbon chains of fatty acids. The absence of delta-12 and delta-15 *desaturase* enzymes in humans is the responsible factor, so for such essential fatty acids humans are dependent on dietary supplements (Nguyen et al., 2019). Two EFAs specifically, omega-3 fatty acids and omega-6 fatty acids, are truly needed by humans in a specific ratio to sustain healthy life by way of active physiological and biological functionality (Alagawany et al., 2019).

The term "bioavailability" relates to a process by which any nutrient is transported to and absorbed at the required site of action for its metabolic fate in a body (Schuchardt and Hann, 2013). A nutrient's essentiality lies in the specific need of the body to perform specific physiological functions. Key steps in making nutrients bioavailable include the following:

- Digestion
- Absorption
- Nutrient circulation by blood
- Assimilation

According to Ghasemifard et al. (2015) the concept of bioavailability is related to that of digestibility. Bioavailability can be understood as the rate at which the substance is absorbed from the gastrointestinal tract and is extensively based on the substance reaching the portal system. Absorption of the substance occurs in the gastrointestinal tract only to a certain extent, depending on the metabolic circumstances. It is worth stating that not all the absorbed nutrients reach the systemic circulation or tissue compartments; there are also other physiological destinations. So, evidence of quantitative differences in the bioavailability of ingested nutrients is vital from the point of view of pharmacokinetics and of dietary planning too.

8.2 METABOLISM

The digestive tract directs the fat to the stomach; minute droplets are then formed by mechanical contraction and relaxation of peristalsis in the stomach. The stream of flow in the upper small intestine section further minces it in preparation for emulsification by bile acids. The resultant emulsified fats are an important prerequisite for *lipases* to cleave triglyceride (TG)-bound fatty acid (FA). *Pancreatic lipase* hydrolyzes TG ester bonds in the small intestine (Figure 8.1). The performance of the aforesaid enzyme in cleaving fatty acid from the *sn*–1/3 position to form 2-monoacyl glycerol (2-MAG) and small amounts of diacyl glycerol (DAG) is advantageous to body. Released free fatty acid (FFA) and 2-MAG are then occupied by enterocytes as mixed micelles via passive diffusion. Fatty acid transport protein (FATP) is also involved here to assist in fatty acid transportation (Stumpf, 1969). Although in vitro studies have shown that reduced FATP4 expression in enterocytes reduced the uptake of long-chain (LC) PUFAs by 50%, the exact role in the intestinal uptake of LC n-3 FA remains unclear.

The transport of fatty acid is initiated with lymph and blood. Primarily, fatty acids are re-esterified to triglycerides in the endoplasmic reticulum followed by their amendment into acyl-CoA via *acyl-CoA synthetase*, and the fatty acids are also esterified later with glycerol to 2-MAG or DAG. Subsequently, rTG, along with cholesteryl esters and plasma lipids (PL), is associated with apolipoprotein B_{48} and other apolipoproteins to form chylomicrons, which are released into the lymph vessels through the basolateral membrane via exocytosis and then enter the bloodstream via the thoracic duct.

LC n-3 FA can be detected in the blood in various forms such as FFA bound to TG and PL, which in turn are integrated as lipoproteinaceous mass. Ultimately, blood transports LC n-3 FA to the target tissues, viz. heart muscle cells, the brain/nervous system and retina cells (Figures 8.1 and 8.2).

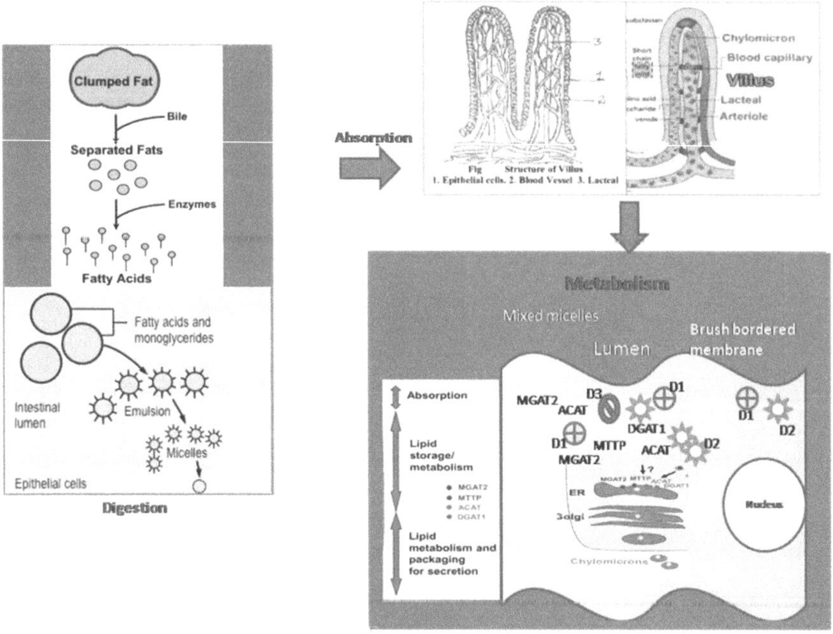

Figure 8.1 Omega fatty acid route from food to tissue (adapted from Schuchardt and Hahn, 2013). Fat emulsified in the stomach is then directed to the small intestine, where the n-3 FAs are cleaved off from their various types of bonds to form free fatty acids and 2-MAG. Free n-3 FA and 2-MAG are taken up as mixed micelles. In the enterocytes, n-3 FAs are re-esterified to triacylglycerides, which are then incorporated into chylomicrons and transferred via the basolateral membrane to the lymph and thus to the systemic circulation. The blood then transports n-3 FAs to the target tissues, where they are primarily incorporated in membranes.

8.3 BIOAVAILABILITY

Different food processing strategies and the presence of "antinutrients" can reduce the bioavailability of many nutrients from food. In some cases, the antinutritive qualities of the ingested material are so predominant in comparison to its nutritive value that its role in the body is altered to that of a threat or toxin. Most of the reduction of nutrients occurs through binding and by degradation means, which restricts them from being digested or absorbed in the body. The bioavailability of long-chain omega-3 polyunsaturated fatty acids depends on various factors such as chemical structure, intake of other food, absence and presence

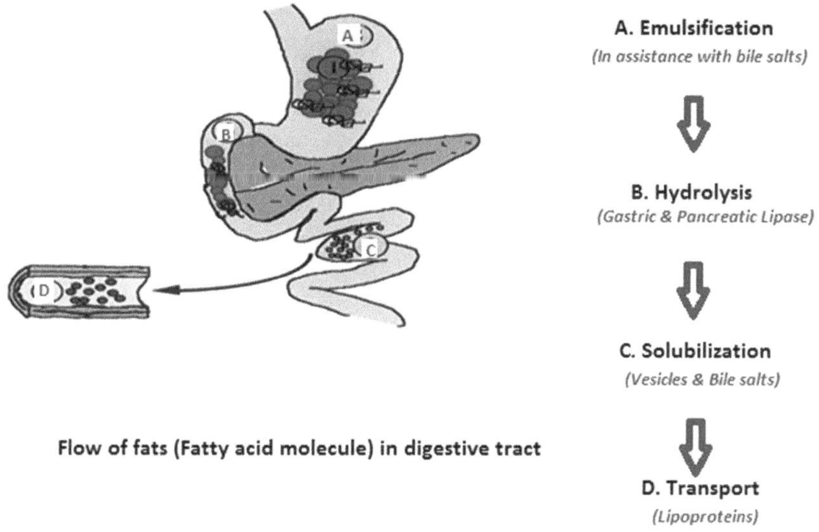

Figure 8.2 Bioavailability route access for fatty acids (adapted from Fave et al., 2004)

of other biomolecules, etc. (Schuchardt and Hahn, 2013). However, digestive mechanism, matrix effects and galenic effects (emulsification and micro-encapsulation) can never be ignored. Furthermore, the presence of lignan, which has been reported to compete with ω-fatty acid digestion, also ceases their bioavailability.

8.4 FACTORS AFFECTING BIOAVAILABILITY OF ESSENTIAL FATTY ACIDS

8.4.1 Chemical Bonds Affecting Bioavailability

LC n-3 fatty acids are the omega fatty acid constituents that have been isolated from fish sources and trans-esterified to oils with modified properties to broaden the end food use (Cholewski et al., 2018). They also find their way, by means of supplementation, enrichment and fortification, into diets, albeit to a much lesser extent. Krill oil, even in the presence of high phospholipids (44%), shows exceptionally elevated bioavailability among the fish oils. In light of this, it is surprising that research to date has focused predominantly on ω-3 fatty acid triglycerides (FA-TG) and long-chain ω-3 fatty acid ethyl esters (FA-EE), whereas very few studies have focused on the availability from oil.

Most investigations have varied in terms of conditions like dose, number of cases and study duration, often making it hard to compare them critically. However, there is more certainty about the effects of triglyceride (TG) attached ω-3 fatty acids, more specifically r-triglyceride (rTG)-bounded fatty acids. In a two-week, short-term study to evaluate various fish oils, rTG showed the best availability compared with non-chemically modified fish oil, FFA and cod liver oil. Some kinetic studies found FFA to have the best bioavailability when compared with ω-3 such as as rTG or EE. However, FFAs are highly susceptible to oxidation and are a cause of gastrointestinal side-effects with fish oils (Loftsson et al., 2016). Therefore, they are usually separated out during the deacidification stages of refining.

Numerous studies have described very significant changes attributable to the bioavailability of ω-3 FA-EE and -TG. In vitro studies prove the inefficient mechanism of *pancreatic lipases* to hydrolyze FA-EE-ethanol bonds; this hydrolyzation occurs 10–50 times less efficiently than that between FA-glycerol bonds in TG. The LC n-3 FA-EE bonds are cleaved by *carboxyl ester lipases*, which appear to be less active than lipases.

8.4.2 Matrix Effect on Bioavailability

It is of concern to all that the fat matrix within food influences the absorption of ω-3 acids ("the matrix effect"); this impacts the introduction, recommendation, and consumption of fat-based formulations impregnated with bioavailable nutrients. Fave et al. (2004) explore the difficulties in obtaining desired cardio-protective effects in the German population by supplementing breakfast with omega-3 PUFAs in competition with the PUFA-deficient diet typically consumed by inhabitants of Germany. Correspondingly, Americans, with an eating style that consists of low PUFAs in breakfast, in total consume just 16% fat throughout the day. In a different view, it is not always true that high-fat foods are very high in omega-3 fatty acids; for example, *Entandrophragma angolense* yields a high proportion of fat but is lacking in essential omega-3 fatty acids. Its seed oil comprises 33.29% oleic (w/w) acid and only 0.2% (w/w) alpha-linolenic acid. Similarly, chia seed oil yielded 27 g/100 g of seeds, of which 64.04% is ALA and only 14.98% are saturated fatty acids (Schuchardt and Hahn, 2013). In a nutshell, high linoleic acid and low omega-6/omega-3 ratio make these foods of dubious nutritional value, especially due to their induction of obesity and other heart-associated risks with few if any mitigating benefits.

8.4.3 Long-Chain Polyunsaturated Fatty Acids and Interactions

Linoleic acid (LA), an ω-6 fatty acid, sometimes works as a precursor and hence competes for its conversion to ω-linolenic acid to ALA, EPA and DHA. The EPA and DHA can be transformed into anti-inflammatory agents. This change in the supply of ω-6 PUFAs to ω-3 fatty acids influences the plasma concentration and is applicable both to non-esterified ω-3 fatty acids and also to those associated with phospholipids, triglycerides and cholesteryl esters (Wood et al., 2014).

Evidence of a reduced supply of ω-6 PUFAs for 12 weeks shows a marked increase in ω-3 PUFA concentration in plasma (Cicero et al., 2015). Wood et al.'s (2014) conclusion reveals that increases in EPA and DHA probably are achieved through reduced supply of linoleic acid and/or increased ALA intake, as routinely seen in diets; the suggested ratio of omega-6/omega-3 is in the range of 1:5 to 1:10, and in Japan an ideal range of 1:2 to 1:4 is commonly found, but the ratio for breastfeeding mothers should not exceed 1:10. Researchers are investigating the legitimacy of returning to an "ancestral diet," which is characterized by the balanced consumption of both groups of fatty acids at the level of 1:1. Attention is also being paid to certain health benefits, such as nervous system maintenance and reduction in tumor cell proliferation, that may result from a strict omega-6/omega-3 ratio of 2.5:1 in the diet.

8.4.4 Role of Lipid Oxidation on Bioavailability

ω-3 fatty acids are prone to oxidation. A few reports have stated that an increase in the consumption of peroxides in fish oil capsules affects health negatively to the same extent as it contributes positively (Ottestad et al., 2016). To sum up these losses, minimization practices for oxidative damage are underway; among these, microencapsulation of oils using whey protein isolate-gum Arabic (WPI-GA) complex coacervates and spray-dried microcapsules are on the front line. Some other lesser-used applications, including ionic gelation by way of a combination of *Salvia hispanica* mucus with sodium alginate and calcium, are also being used to make them shelf stable. Staprans et al. (1994) postulate that the oxidative loss and absence of bioavailability upon consumption of peroxidized vegetable oils is due to the increase in the number of peroxidized lipids in the chylomicrons. In another report by Ottestad et al. (2016), consumption of oxidized cod liver oil had no impact on triacylglyceride-rich lipoprotein levels, even when ω-3 fatty acids were incorporated into the oil (Table 8.1).

Table 8.1 Studies Addressing the Bioavailability of Essential Fatty Acids

Study	References
The bioavailability and pharmacodynamics of different concentrations of omega-3 acid ethyl esters	Bryhn et al., 2006
Bioavailability of marine n-3 fatty acid formulations	Dyerberg et al., 2010
Long-term effects of DHA-bound phospholipids and the combination of DHA-bound triglyceride and egg yolk phospholipid on lipid metabolism in mice	Che et al., 2018
Effects of microencapsulation on bioavailability of fish oil omega-3 fatty acids	Christopherson et al., 2016
Omega-3 long-chain fatty acid "bioavailability": A review of evidence and methodological considerations	Ghasemifard et al., 2014
A review of evidence and methodological considerations	Ghasemifard et al., 2015
The bioavailability of an omega-3-rich algal oil is improved by nanoemulsion technology using yogurt as a food vehicle	Lane et al., 2014b
Bioavailability of food active compounds: A challenging journey to bioefficacy	Rein, 2013
Bioavailability of eicosapentaenoic and docosahexaenoic n-3 polyunsaturated fatty acids in salmon patties compared with capsules	Maki et al., 2003
Strategies to improve bioavailability of omega-3 fatty acids from ethyl ester concentrates	Maki and Dicklin, 2019
Effects of a self-micro-emulsifying delivery system formulation versus a standard ω-3 acid ethyl ester product on the bioavailability of EPA and DHA: A study in healthy men and women in a fasted state	Maki et al., 2018
Bioavailability of fatty acids from krill oil, krill meal and fish oil in healthy subjects–a randomized, single-dose, cross-over trial	Kohler et al., 2015
A single-dose, comparative bioavailability study of a formulation containing OM3 as phospholipid and free fatty acid to an ethyl ester formulation in the fasting and fed states	Laponite et al., 2019

(*Continued*)

Table 8.1 (Continued) Studies Addressing the Bioavailability of Essential Fatty Acids

Study	References
Dose effect of ALA on PUFA conversion, bioavailability and storage in the hamster	Morise et al., 2004
Bioavailability and potential uses of vegetarian sources of omega-3 fatty acids: A review of the literature	Lane et al., 2014a
Bioavailability of long-chain omega-3 fatty acids	Schuchardt and Haan, 2013
Dietary intake of n-3 PUFAs modifies the absorption, distribution and bioavailability of fatty acids in the mouse gastrointestinal tract	Yang et al., 2017
Formulation strategies to improve the bioavailability of poorly absorbed drugs with special emphasis on self-emulsifying systems	Gupta et al., 2013
Bioavailability of dietary omega-3 fatty acids added to a variety of sausages in healthy individuals	Kohler et al., 2017
A comprehensive review of chemistry, sources and bioavailability of omega-3 fatty acids	Cholewski et al., 2018
Bioavailability of omega-3 essential fatty acids from perilla seed oil	Kurowska et al., 2003
Steady-state bioavailability of prescription omega-3 on a low-fat diet is significantly improved with a free fatty acid formulation compared with an ethyl ester formulation: The ECLIPSE II study	Offman et al., 2013
Systemic bioavailability and dose proportionality of omega-3 administered in free fatty acid form compared with ethyl ester form: Results of phase 1 study in healthy volunteers.	Offman et al., 2017
Bioavailability of n-3 fatty acids from n-3-enriched foods and fish oil with different oxidative quality in healthy human subjects: A randomized single-meal cross-over study	Ottestad et al., 2016
Long-chain omega-3 FA bioavailability: Implications for understanding the effects of supplementation on heart disease risk	Maki et al., 2018

(Continued)

Table 8.1 (Continued) Studies Addressing the Bioavailability of Essential Fatty Acids

Study	References
Oil nanoencapsulation: Development, application, and incorporation into the food market	Ferreira and Nunes, 2019
Comparative bioavailability of dietary ALA and DHA in growing rat	Ballihaut et al., 2001
Bioavailability of long-chain omega-3 polyunsaturated fatty acids from phospholipid-rich herring roe oil in men and women with mildly elevated triacylglycerols	Cook et al., 2016
A novel omega-3 free fatty acid formulation has dramatically improved bioavailability during a low-fat diet compared with omega-3-acid ethyl esters: The ECLIPSE (EpanovaW compared to LovazaW in a pharmacokinetic single-dose evaluation) study	Davidson et al., 2012
What is the most effective way of increasing the bioavailability of dietary long-chain omega-3 fatty acids—daily vs. weekly administration of fish oil?	Ghasemifard et al., 2015
Orally administered [14C]DPA and [14C]DHA are metabolized differently to [14C] EPA in rats	Kaur et al., 2013
Food microstructure affects the bioavailability of several nutrients	Prada and Aguilera, 2007
Comparison of bioavailability of krill oil versus fish oil and health effect	Ulven and Holven, 2015
Bioavailability of ALA from flaxseed diets as a function of the age of the subject	Patenaude et al., 2009
Bioavailability of ALA in subjects after ingestion of three different forms of flaxseed	Austria et al., 2008
Bread enriched with microencapsulated tuna oil increases plasma docosahexaenoic acid and total omega-3 fatty acids in humans	Yep et al., 2002
Krill oil: Evidence of a new source of PUFA with high bioavailability	Cicero and Colletti, 2015
A novel self-micro-emulsifying delivery system (SMEDS) formulation significantly improves the fasting absorption of EPA and DHA from a single dose of an omega-3 ethyl ester concentrate	Qin et al., 2017

(Continued)

Table 8.1 (Continued) Studies Addressing the Bioavailability of Essential Fatty Acids

Study	References
A Novel omega-3 acid ethyl ester formulation incorporating Advanced Lipid Technologies™ (ALT®) improves docosahexaenoic acid and eicosapentaenoic acid bioavailability compared with Lovaza	Lopez-Toledano et al., 2017
Advanced Lipid Technologies (ALT): A proven formulation platform to enhance the bioavailability of lipophilic compounds	Lopez-Toledano et al., 2019
Dietary omega-6 fatty acid lowering increases bioavailability of omega-3 polyunsaturated fatty acids in human plasma lipid pools	Taha et al., 2014
Bioavailability of n-3 polyunsaturated fatty acids (PUFA) in foods enriched with microencapsulated fish oil	Wallace et al., 2000
The bioavailability of eicosapentaenoic acid from reconstituted triglyceride fish oil is higher than that obtained from the triglyceride and monoglyceride forms	Wakil et al., 2010b
Marine lipid-based liposomes increase in vivo FA bioavailability	Cansell et al., 2003
Incorporation of EPA and DHA into plasma phospholipids in response to different omega-3 fatty acid formulations—a comparative bioavailability study of fish oil vs. krill oil	Schuchardt et al., 2011
Enhanced increase of omega-3 index in response to long-term n-3 FA supplementation from triacylglycerides versus ethyl esters	Neubronner et al., 2011

8.4.5 Lipid Droplet Size

Lipids are aggregates of variously sized droplets and are available from animal and plant sources, e.g., milk products and vegetable oils, respectively. Besides this, many emulsion and aqueous organic biphasic systems also operate in the same way. The lipid droplet sizes are directly related to the extent of fat emulsification according to its lipid/water interfacial interactions. So as the interfacial tension is reduced, interaction increases, which in turn raises the bioavailability (Fave et al., 2004).

Lipid droplet organization also has an impact on the bioavailability features of fatty acids, as seen in milk fat composition. A lipid globule of milk fat consists of a hydrophobic core mostly of cholesterol, triglycerides and fat-soluble vitamins. On the outside, these globules are sheeted with phospholipids (amphipatic) that allow these droplets to maintain solubility and stability in an aqueous environment. Certainly, some fatty constituents, viz. cholesterol and triglycerides in the outer layer, impact bioavailability through the high hydrolyzing capacity of lipases and other acids (Brockman, 1984).

8.4.6 Role of Aging in Bioavailability

Age also appears to influence bioavailability access for EPA and DHA. An intervention study reveals an increase in the DHA level in plasma of healthy older subjects (74 ± 4 years) compared to younger subjects (24 ± 2 years). The administration of 1 g EPA and DHA per day over a period of 3 weeks surprisingly showed more of an increase in DHA in the older group than in the younger subjects. This was presumably due to different levels of assimilation of DHA, which varied from person to person on an age fraction basis, but no such observations were made for EPA's impact on plasma lipid profiles of aged individuals (Yehuda et al., 2005).

8.4.7 Parenteral Administration

It is difficult to postulate any hypothesis with full validation for the parenteral administration of omega-3 fatty acids in relation to the healthy population. This is because the supply is mainly reserved for patients (adults and preterm infants) undergoing intensive therapy. In addition, it is worth noting that parenteral administration of mixtures based on fish oil is drastic, in that it may lead to biochemical liver damage and even progression of fibrosis in organs (Martin, 2015).

Clinical outcomes for 20 patients ready for surgical elimination of colorectal metastasis from liver cancer were investigated by Al-Taan et al. (2013). Medically, they were examined for liver function and plasma lipid level within the reference range, keeping in mind the aim of study: to access fatty acids in plasma phosphatidylcholine (PC) and erythrocytes during and after the intravenous infusion of oil emulsion. PC phospholipid was detected at all times in the circulatory blood and erythrocyte (membrane) during and shortly after the intravenous infusion of the oil emulsion. Parenteral administration of DHA and EPA lipid emulsion allowed a rapid and significant increase in their blood levels (EPA/DHA in plasma PC and EPA in erythrocytes). However, EPA levels returned to their initial values within 5–12 days after the end of the infusion. The study has directed researchers' view toward achieving quicker turnover of EPA than DHA in cells whether supplied by oral or intravenous mode and at any moment.

8.5 STRATEGIES OF MITIGATION WITH IMPROVED BIOAVAILABILITY CHARACTERISTICS

8.5.1 Emulsification as a Tool

Fat-rich diets excite the pancreas to secrete fat-hydrolyzing enzymes, cause the gallbladder to eject bile for emulsification purposes and activate *pancreatic lipase*. However, it is recommended that persons with high cardiovascular risk should reduce their supply of animal fat. On the other hand, substituting animal fat with vegetable fat is not necessarily a good solution, taking into account the low content of omega-3 fatty acids in this substituted diet. Nanotechnological advancements, viz. nanoparticles, nanospheres, nanocapsules, nanoemulsions and self-emulsifying drug delivery systems (SEDDS), significantly recover the absorption of fatty acids in nanoparticle ranges (McClements et al., 2008).

The self-microemulsifying delivery system (SMEDS) for ω-3 fatty acid ethyl esters and other emulsified compounds in stomach is easily incorporated into body tissues without any need for a high-fat meal. It has been known for many years that emulsified omega-3 acids are characterized by equally good bioavailability in low- and high-fat environments. Qin et al. (2017) also demonstrate that the absorption of DHA + EPA contained in a similar system (SMEDS) was 6 times higher than that of DHA + EPA alone. The effective use of SMEDS technology can be visualized along with long-term research plans for dramatic structural modulations of specific cell membranes, especially erythrocytes.

The absorption of omega-3 fatty acid ethyl esters can also be enhanced through the use of Advanced Lipid Technologies (ALT). ALT is a special lipophilic system that, regardless of the supply of food and the fat content thereof, increases the bioavailability of lipid-based compounds, including omega-3 fatty acid ethyl esters, generating the spontaneous formation of micelles. An example of a preparation equipped with this component is SC401 (DHA and EPA ethyl esters + Advanced Lipid Technologies). In the supply of low-fat food, the intake of SC401 was associated with significantly higher values demonstrating the bioavailability of DHA and EPA than those seen for Lovaza (a medicine used to treat inflammation of the pancreas).

Free fatty acids are another example of a formula well absorbed with a low fat content in the diet. Epanova bioavailability in conditions of low fat supply is much greater than with Lovaza (AUC_{0-t} for Epanova is four times higher than for Lovaza); consumed with high-fat food, it has similar bioavailability as SC401 accepted in conditions of low food supply. Vegetarians and vegans will also be able to use the emulsification technique in line with their own lifestyle, as the results of the latest research indicate the possibility of using vegetable proteins from pea, lentil and faba bean as emulsifiers. Moreover, oil-in-water nanoemulsions are constructed using vegetarian oils (algae and flaxseed oil).

A very common method for the incorporation of nanoemulsions into food is its transformation, most often into the form of powder, using microencapsulation (Walker et al., 2015). Microencapsulation stabilizes the oil mixture, protects it against oxidation and eliminates the phase difference often present when mixed with food. The microencapsulation of fish oil increases the bioavailability of omega-3 PUFAs to a value very close to the bioavailability of these acids in meals rich in fish oil in liquid form. Emulsification increases the bioavailability of LC omega-3 PUFAs by increasing the efficiency of incorporation into triacylglyceride-rich lipoproteins. Emulsifiers also modify the expression of genes responsible for the transport of fatty acids in enterocytes. Emulsification (pre-emulsification) increases the absorption of LC PUFAs, including DHA, EPA and ALA, but does not affect the absorption of fatty acids having shorter chain lengths and fewer unsaturated bonds.

Of course, emulsification elevates the bioavailability of omega-3 acids not only in the blood, but also in the lymph. Emulsifiers used in food are part of this mechanism. For example, soya lecithin added to flaxseed oil increases the number and size of ALA-rich chylomicrons produced by enterocytes. However, another emulsifier, sodium caseinate, significantly reduces the absorption of ALA in the intestine. This is probably related to the effect on the expression of fatty acid binding protein-2 (FABP2) gene, which is involved in the transport of fatty acids in enterocytes. It was noted that the presence of soy lecithin in the emulsion was accompanied by high expression of this gene, whereas the use of sodium caseinate was associated with a decrease in FABP2 gene expression. This process requires careful research in humans, because it is likely that the effect of emulsifiers on gene expression should be considered during food production, especially when we expect a specific healing effect.

Maki et al. (2018) discussed an absorption enhancement technology, i.e., a self micro-emulsifying delivery system (SMEDS), as one of the strategies to enhance the absorption of long-chain omega-3 polyunsaturated fatty acids. By enabling rapid emulsification and micro-droplet formation after ingestion in the gut under aqueous conditions, it enhances the digestion and absorption of omega-3 fatty acids (EPA and DHA) in the absence of bile. Adoption of this technique enhances the lipophilic agent absorption without any additional fat requirement/intake. Finally, it is concluded that the bioavailability of DHA and EPA rose through SMEDS formulation even when administered in a fasting state. Earlier, Qin et al. (2017) concluded that the absorption of EPA and DHA as ethyl ester in the SMEDS formulation was recorded to be six times greater than that of omega-3-acid EE oil.

8.5.2 Encapsulation

Encapsulation is an efficient delivery technique for trapping desired substances in oil. Microencapsulation technology is one of the strategies adopted by the food industry to protect PUFAs from oxidation to preserve long-chain omega-3 polyunsaturated fatty acids during processing and storage (Yep et al., 2002). Milk proteins such as whey proteins, when used as wall material for encapsulation, were reported to enhance the bioavailability of microencapsulated omega-3 fatty acids (Kaushik et al., 2015). Several methodologies have been employed for the production of nanocapsules containing oils, such as emulsification-solvent evaporation, spontaneous emulsification, emulsion-diffusion, high-shear emulsification, homogenization, spray drying, emulsion and supercritical fluid extraction process, etc.

Wakil et al. (2010a) prove that microencapsulation also enhances availability features of FA as demonstrated with emptied pollen exine hollow shells filled with dried fish oil powder as the encapsulated material.

8.5.3 Effect of Galenic Formulation on Bioavailability

So far, a few studies have examined the influence of the galenic formulation of fish oil supplements on the bioavailability of LC n-3 FA. Changes to the galenic formulation of an LC n-3 FA supplement generally aim to improve tolerance and/or increase availability. Two kinetic studies showed that the consumption of LC n-3 FA oils was significantly less effective compared to the consumption of the same LC n-3 FA oils in an emulsified form, which produced a more pronounced increase in EPA and DHA in PL (Schuchardt and Hahn, 2013). The higher availability of emulsified LC n-3 FA oils is probably due to finely dispersed oil droplets, which enable *pancreatic lipases* to cleave TG-bound LC n-3 FA more easily. The dispersion of fat globules is an elementary process of gastrointestinal lipid digestion and a prerequisite for *hydrophilic lipases* to cleave lipids. Ingesting a lipid emulsion containing LC n-3 FA simplifies this physiological dispersion, resulting in better digestion and absorption of LC n-3 FA.

A recent study by Sanguansri et al. (2013) used patients with a permanent ileostomy (artificial ileum ostomy) to investigate differences in LC n-3 FA bioavailability between different food products (orange juice, yogurt and cereal bar) fortified with microencapsulated fish oil. Patients (*n*=6) in this kinetic study received alternately 284 mg LC n-3 FA (50 mg EPA and 234 mg DHA) via different food products or 266 mg LC n-3 FA (48 mg EPA and 218 mg DHA) via fish oil capsules in a cross-over manner. Following ingestion of LC n-3 FA, the amount of undigested or unabsorbed LC n-3 FA was measured in the ileal effluent after 0, 2, 4, 6, 8, 10, 12, 14, 16 and 24 h in order to calculate the amount of released, digested and absorbed LC n-3 FA. The study showed that the bioavailability levels between different food products fortified with microencapsulated fish oil and fish oil gelatin capsules were equal. However, the study had a number of potential limitations: for example, the degradation of LC n-3 FA by digestive enzymes produced by the body cannot be distinguished from that due to bacteria. It is incomprehensible why the authors abstained from collecting blood samples and analyzing the amount of LC n-3 FA absorbed, which could have served as a second measure for the evaluation of the bioavailability.

The increasingly common practice of applying an enteric coating (which is resistant to gastric acid) to fish oil capsules appears to have no impact on bioavailability. Two studies have been conducted on this topic so far. The authors used EPA- and DHA-rich fish oil capsules and ALA-rich perilla seed oil capsules with enteric coatings and compared them with regular soft gelatin capsules containing the same oil, respectively. The results showed no major differences in bioavailability. An investigation of Schneider et al. (2011) reported a slightly delayed increase in EPA and DHA plasma lipid levels as a bioavailability concern for ingestible capsules developed with an enteric coating. This time-shifted uptake could be due to a delayed digestion of lipids, more precisely the hydrolysis of TG. The digestion of lipids starts with the hydrolysis of TG to 2-MAG and FFA catalyzed by gastric and sublingual *lipases*. The delayed release of fish oil from the coated capsules might also result in a delayed hydrolysis of TG-bound FA. However, the major part of lipids is digested by pancreatic *lipases* in the upper segment of the jejunum. Due to the relatively long duration of lipids in this segment, the incubation time of *pancreatic lipases* is probably long enough to uniformly hydrolyze EPA and DHA TG from coated and uncoated capsules. Similarly, that is probably the reason why the absolute absorption – as measured by the AUC_{0-72} values – was similar for the two capsule forms.

8.5.4 Enzymatic Pre-treatments Affecting Bioavailability

Metabolic interactions of the omega-3 and omega-6 fatty acids occur in cellular lipids and plasma with consumption of a diet that results in the production of inositol phosphates, which influence the production of lipid mediators and hormone-like biochemical compounds known as eicosanoids (Kinsella et al., 1990). As aforementioned, most dietary fatty acids are consumed in the form of triglycerides, an ester formed from glycerol and three fatty acid groups. After omega-3 and omega-6 fatty acids are consumed, they undergo various metabolic processes such as beta-oxidation, esterification, conversion to longer chain, desaturation, elongation, etc. mainly in liver and in some cases other tissues also. In the case of the human digestive tract, the process of de-esterification is performed mainly by *human gastric lipases* (HGL) and *human pancreatic lipases* (HPL). The formation of a lipid/water interface finally determines the bioavailability of dietary fats (Hedge et al., 2020).

In the human digestive tract, water-soluble human pancreatic lipases intermingled through active sites for large fat globules. Bile salts get released in the bile duct as soon as dietary fat is detected. Bile salts play a significant role in the digestion of fats, which undergo a process known as emulsification, in which large fat globules are broken into smaller droplets in the small intestine. Emulsification greatly increases the surface area of the lipid globules, which allows greater access of human pancreatic lipases to the lipids and improves the de-esterification efficiency after hydrolysis of either triglycerides or ethyl ester. The resulting monoglycerides and free fatty acids get associated with bile salts and phospholipids finally to form micelles. The lipolytic products are transported primarily via micelles through the mucous layer and absorbed by the microvilli. Long-chain fatty acids are re-synthesized in the enterocytes to triglycerides and phospholipids and packed into chylomicrons for transportation via the lymphatic system to the systemic circulation (Figure 8.3).

A-linolenic acid (ALA, 18:3n-3)

⇓ *Δ 6-desaturase*

Stearidonic acid (SDA, 18:4n-3)

⇓ *Elongase*

Eicosatetraenoic acid (ETA, 20:5n-3)

⇓ *Δ 5-desaturase*

Eicosapentaenoic acid (EPA, 20:5n-3)

⇓ *Elongase*

Docosapentaenoic acid (DPA, 22:5n-3) *Elongase* ⇒ Teracosapentaenoic acid (24: 5n-3)

⇓ *Δ 4-desaturase*

Docosahexaenoic acid (DHA, 22:6n-3) *Δ 6-desaturase* ⇒ Tetracosahexaenoic (24:6n-3) acid

Figure 8.3 Biosynthesis and metabolic pathways of omega-3 polyunsaturated fatty acid (adapted from Nguyen et al., 2019)

8.6 STRATEGIES TO IMPROVE EFA BIOAVAILABILITY OF DIFFERENT FOODS

8.6.1 Milk and Meat Products

Nutritional manipulation such as the supplementation of animal feed with vegetable oils and seed oils slightly increased EPA, DHA and DPA content in milk in both bovines and ovines, as reported by Khas et al. (2010). Feeding dairy animals with a marine oil–enriched diet was shown to produce high concentrations of long-chain omega-3 fatty acids such as EPA and DHA in their milk and milk products (Nguyen et al., 2019). These observed effects were achieved through diet-shifting behavior to raise the fatty acid profile in both meat and milk products with supplemented or fortified conventional grain, grass and legume forage. This transformation in animal diet indicates the existence of nutritional variation among organically and conventionally based production systems (Benbrook et al., 2018), supporting the eco-friendly concept of practicing sustainable agriculture.

8.6.2 Poultry and Eggs

The above-described dietary adjustments to enhance fatty acid bioavailability are similarly applicable for the case of poultry. Here, omega-3 fatty acids can be introduced to the body through the concept of designer eggs (Ehr et al., 2017). These eggs are considered to be enriched in omega-3 fatty acid levels and hence contribute to the improvement of human health by promoting a balanced omega-6:omega-3 ratio. In general, consumers are benefitted when there is more than 600 mg of long-chain omega-3 fatty acid bioavailability in eggs (Alagawany et al., 2018). The achievement of increased ω-3 fatty acid levels in eggs can be enhanced by supplementing the hen's diet with special dietary supplements including fish oil, fish meal, wheat soybean meal, safflower oil, pomegranate seed oil, groundnut oil, linseed oil, hemp oil, hemp seed, alfalfa sprouts and some algal species such as *Phaeodactylumtricornutum, Nannochloropsisoculata, Isochrysisgalbana, Chlorella fusca,* etc. in them. Such supplementation also influences meat quality, fertility characteristics and egg laying frequency in hens (Alagawany et al., 2019).

8.6.3 Algae (Phytoplankton and Zooplankton)

Use of oil extracted from marine algal sources is one of the best remedies for the vegetarian population. Instead of the use of kosher (rules for the preparation, sale and consumption of food as per Jewish law) and halal (meat preparation as per Muslim law) products by vegans, the use of DHA-fortified algal oil is a revolutionary step; this also could be commercially used to feed the human population solely with vegetarian sources (Lane et al., 2014a).

Algal diets have been investigated to be used as rich omega-3 sources, as both EPA and DHA are naturally contained within them. The main focus in acquiring EPA and DHA from algae should be on the lipid structure by which EPA and DHA are ingested, since this will influence bioavailability and accumulation. Recently, an interesting and promising study has succeeded in extracting high EPA levels from diatoms such as *Phaeodactylum tricornutum,* which seems less expensive than that from cod liver. Similarly, DHA-rich oil content has been produced from microalgae and *Ulkenia* sp. Some algae species and dinoflagellates appear as potentially rich sources of n-3 omega PUFAs in the very near future; this will all be possible because of the ease of using phytoplankton and zooplankton.

Microalgae are one of the primary producers of long-chain omega-3 polyunsaturated fatty acids, mainly EPA and DHA. Among the numerous algal species cited here that have been identified as rich in DHA, EPA and DHA/EPA together are *Chlorella minutissima, Nannochloropsis* sp., *Nannochloropsis oceanica, Nannochloropsis salina, Pinguiococcus pyrenoidosus, Dunaliella salina, Crypthecodinium cohnii, Ceratium horridum, Phaeodactylum tricornutum, Pavlova lutheri* and *Isochrysis galbana* (Amjad Khan et al., 2017).

As per the study of Lane et al. (2014b), algal species (*Crypthecodinium cohnii* and *Schizochytrium* spp.) are two major sources of DHA, with total fatty acid levels of 55% and 40%, respectively. ω-3 PUFAs (EPA and DHA) are synthesized by phytoplankton and algae and then eventually transferred via the food chain to fish and other marine creatures. Thraustochytrids, which are isosmoheterotrophic, unicellular monocentric fungi-like protists, can grow heterotrophically and can also synthesize high amounts of DHA. Similarly *Aurantiochytrium* species is a promising candidate for commercial production of ω-3 fatty acids such as DHA (Lane et al., 2014b).

Excessive use of fish oils such as DHA and EPA by the global human community is increasing as the human population is becoming more aware of their health benefits. The fish industries have encountered high demand and low supply, and thus are unable to satisfy consumer needs at the global level. In order to bridge the gap between current intake, 100–200 mg/day, and recommended levels, 650 mg/day, policies should be enacted to cater to the population through technological enrichment of a variety of foods with EPA/DHA as a suitable remedy (Ganesan et al., 2014).

The correlation between ω-3 polyunsaturated fatty acids and the metabolism of hormones such as catecholamine (adrenalin and noradrenalin) needs thorough investigation, as higher chances of neurological disorders such as schizophrenia have been associated with their consumption.

8.6.4 Organic Agriculture

The adoption of the concept of sustainable agriculture through organic practices, now accepted all over the globe, is aimed at producing agro-based products with less apprehension about environmental and nutritional security. High amounts of ω-3 fatty acids such as alpha-linolenic acid (ALA) and eicosapentaenoic acid (EPA) are reported to occur in organic milk in comparison with conventional milk. A balanced omega-6:omega-3 ratio was recorded in organic milk produced using organic production systems; however, climatic conditions also influence the fatty acid composition of milk with variation in results, as the supplementation of diet varies between winter and summer (Muller et al., 2017).

8.6.5 Alpha-Tocopheryl Phosphate Mixture (TPM)

Alpha-tocopheryl phosphate mixture (TPM) is a safe form to encapsulate and solubilize materials that have poor solubility features. TPM increases the in vitro solubility of a model lipophilic nutrient, coenzyme Q10. Various formulations have been employed previously to improve the oral bioavailability of DHA/omega-3. ω-3 DHA formulated with TPM increased the bioavailability of DHA versus control treatments as measured in plasma. TPM can replace or complement the gastric emulsification usually required for efficient oral absorption; however, a lower dose of TPM may result in incomplete DHA emulsification, leading to poor absorption (Libinaki and Gavin, 2017). They concluded that co-administration of omega-3 with TPM enhances the oral bioavailability of DHA in plasma.

Dyerberg et al. (2010) are of the view that rTG of fish oil has superior bioavailability to ethyl ester, which has lower bioavailability, whereas free fatty acids have medium bioavailability. Cholewski et al. (2018) supported the study accordingly, where the bioavailability of omega-3 fatty acids depends on the type of chemical binding, i.e., the structure of lipid: phospholipids > re-esterified triglycerides > triglycerides > free fatty acid > ethyl ester. Recently Laponite et al. (2019) supported the enhanced bioavailability of the free fatty acid form and the phospholipid form compared to the ethyl ester form in a state of fasting. FFA and PL are comparatively more bioavailable than the EE form of EPA and DHA. Several animal studies supported the high bioavailability aspect of phospholipid compared to other forms of omega-3 fatty acids (Ghasemifard et al., 2014).

8.7 STRATEGIES FOR DECREASING FATTY ACID BIOAVAILABILITY

At some point, it may be necessary to decrease fatty acid bioavailability when body lipid storage or blood lipids levels are too elevated, leading to a high risk of cardiovascular diseases, such as in obesity and hyper-lipoproteinemia, respectively. The use of triglyceride structures to impart low digestion and absorption rates can help maintain sufficient fat in the diet of such patients, which would maintain the good taste of food, thus allowing better compliance with the diet. Such triglycerides, called SALATRIM, have been produced by the Nabisco (East Hanover, NJ, USA) and it has been shown that the stearic acid absorption rate decreases markedly when SALATRIM is located at the external positions of the glycerol backbone (Mattson et al., 1979). The mechanism involved is a decrease in the solubilization of stearic acid released by lipases from the glycerol backbone into bile salt–mixed micelles and/or vesicles in the duodenal lumen due to a strong tendency of stearic acid, a long-chain saturated fatty acid, to bind calcium, leading to the formation of calcium soaps that will be eliminated in the feces. These triglycerides represent a family of heterogeneous molecules composed of at least one molecule of stearic acid at the external position of the glycerol and one molecule of short-chain fatty acid (acetic, butyric or propionic acid) and provide only 5 kcal/g instead of 9 kcal/g because of the decreased absorption rate of the fatty acids.

8.8 CONCLUSIONS

In the recent past, essential fatty acids have been explored as a valuable nutritional asset to sustain healthy human life. Present statistics also highlight their value through significant progress in research on bioavailability access for long-chain omega-3 fatty acids. Investigations have claimed that marine-based food consumption as a major source of long-chain omega-3 fatty acids is insufficient for the enhancement of the concentration of long-chain omega-3 fatty acids in the global human diet. They suggest improving other dietary sources too, such as those obtained from animals, like milk, dairy products and eggs. Water pollution is the major threat that drives consumers to demand less marine fish and fish products, as they want to avoid exposure to hazardous pollutants like mercury methyl and other toxins in marine ecosystems. Attempts to increase the content of long-chain omega-3 fatty acid in milk and milk products are considered as one promising target for food technology researchers. The application of technologically advanced research tools such as nanotechnology, genetic engineering, fat tailoring and bioinformatics in the future is also one of the key factors for healthy adaptation. Along with this, understanding of nutritional-genetics is directly needed to explore the metabolism fate of essential fatty acids. The development of bioengineered plant varieties by gene transfer into higher plants (oilseed) from microalgae is an effective strategy to keep pace with society's need for nutritional balance. The current deliverables of use of algal oils, such as krill oil, are among those achievements that prove themselves beneficial to the inhabitants of any geographical territory, since they are safe and have much bioavailable essential fatty acid content. Moreover, there is much unrealized potential in this direction to present the world with newer food products.

REFERENCES

Alagawany, M., Elnesr, S. S., Farag, M. R., Abd El-Hack, M. E., Khafaga, A. F., Taha, A. E. and Dhama, K. (2019). Omega-3 and Omega-6 fatty acids in poultry nutrition: Effect on production performance and health. *Animals: An Open Access Journal from MDPI, 9*(8), 573.

Alagawany, M., Farag, M. R., Dhama, K. and Patra, A. (2018). Nutritional significance and health benefits of designer eggs. *World's Poultry Science Journal, 74*(2), 317–330.

Al-Taan, O., Stephenson, J. A., Spencer, L., Pollard, C., West, A. L., Calder, P. C. and Dennison, A. R. (2013). Changes in plasma and erythrocyte omega-6 and omega-3 fatty acids in response to intravenous supply of omega-3 fatty acids in patients with hepatic colorectal metastases. *Lipids in Health and Disease, 12*(1), 64.

Austria, J. A., Richard, M. N., Chahine, M. N., Edel, A. L., Malcolmson, L. J., Dupasquier, C. M. and Pierce, G. N. (2008). Bioavailability of alpha-linolenic acid in subjects after ingestion of three different forms of flaxseed. *Journal of the American College of Nutrition, 27*(2), 214–221.

Benbrook, C. M., Davis, D. R., Heins, B. J., Latif, M. A., Leifert, C., Peterman, L. and Baranski, M. (2018). Enhancing the fatty acid profile of milk through forage-based rations, with nutrition modeling of diet outcomes. *Food Science and Nutrition, 6*(3), 681–700.

Brockman, H. L. (1984). General features of lipolysis: Reaction scheme, interfacial structure and experimental approaches. In: Lipases, Borgstrom, B. and Brockman, H. L. (eds.), Elsevier, Amsterdam, 3–46.

Bryhn, M., Hansteen, H., Schanche, T. and Aakre, S. E. (2006). The bioavailability and pharmacodynamics of different concentrations of omega-3 acid ethyl esters. *Prostaglandins, Leukotrienes and Essential Fatty Acids, 75*(1), 19–24.

Cansell, M., Nacka, F. and Combe, N. (2003). Marine lipid-based liposomes increase in vivo FA bioavailability. *Lipids, 38*(5), 551.

Che, H., Cui, J., Wen, M., Xu, J., Yanagita, T., Wang, Q. … and Wang, Y. (2018). Long-term effects of docosahexaenoic acid-bound phospholipids and the combination of docosahexaenoic acid-bound triglyceride and egg yolk phospholipid on lipid metabolism in mice. *Journal of Ocean University of China, 17*(2), 392–398.

Cholewski, M., Tomczykowa, M. and Tomczyk, M. (2018). A comprehensive review of chemistry, sources and bioavailability of omega-3 fatty acids. *Nutrients, 10*(11), 1662.

Christophersen, P., Yang, M. and Mu, H. (2016). Effects of microencapsulation on bioavailability of fish oil omega-3 fatty acids. In: Lakkis J. M. (ed.), *Encapsulation and Controlled Release Technologies in Food Systems*, John Wiley & Sons Ltd., UK, 309–332.

Cicero, A. F. and Colletti, A. (2015). Krill oil: Evidence of a new source of polyunsaturated fatty acids with high bioavailability. *Clinical Lipidology, 10*(1), 1–4.

Cicero, A. F., Morbini, M. and Borghi, C. (2015). Do we need 'new' omega-3 polyunsaturated fatty acids formulations. *Expert Opinion on Pharmacotherapy, 16*(3), 285–288.

Cook, C. M., Hallaråker, H., Sæbø, P. C., Innis, S. M., Kelley, K. M., Sanoshy, K. D. and Maki, K. C. (2016). Bioavailability of long chain omega-3 polyunsaturated fatty acids from phospholipid-rich herring roe oil in men and women with mildly elevated triacylglycerols. *Prostaglandins, Leukotrienes and Essential Fatty Acids, 111*, 17–24.

Davidson, M. H., Johnson, J., Rooney, M. W., Kyle, M. L. and Kling, D. F. (2012). A novel omega-3 free fatty acid formulation has dramatically improved bioavailability during a low-fat diet compared with omega-3-acid ethyl esters: The ECLIPSE (Epanova® compared to Lovaza® in a pharmacokinetic single-dose evaluation) study. *Clinical Lipidology, 6*(6), 573–584.

Dyerberg, J., Madsen, P., Møller, J. M., Aardestrup, I. and Schmidt, E. B. (2010). Bioavailability of marine n-3 fatty acid formulations. *Prostaglandins, Leukotrienes and Essential Fatty Acids, 83*(3), 137–141.

Ehr, I. J., Persia, M. E. and Bobeck, E. A. (2017). Comparative omega-3 fatty acid enrichment of egg yolks from first-cycle laying hens fed flaxseed oil or ground flaxseed. *Poultry Science, 96*(6), 1791–1799.

Fave, G., Coste, T. C. and Armand, M. (2004). Physicochemical properties of lipids: New strategies to manage fatty acid bioavailability. *Cellular and Molecular Biology, 50*(7), 815–832.

Ferreira, C. D. and Nunes, I. L. (2019). Oil nanoencapsulation: Development, application, and incorporation into the food market. *Nanoscale Research Letters, 14*(1), 9.

Ganesan, B., Brothersen, C. and McMahon, D. J. (2014). Fortification of foods with omega-3 polyunsaturated fatty acids. *Critical Reviews in Food Science and Nutrition, 54*(1), 98–114.

Ghasemifard, S., Sinclair, A. J., Kaur, G., Lewandowski, P. and Turchini, G. M. (2015). What is the most effective way of increasing the bioavailability of dietary long chain omega-3 fatty acids—Daily vs. weekly administration of fish oil? *Nutrients, 7*(7), 5628–5645.

Ghasemifard, S., Turchini, G. M. and Sinclair, A. J. (2014). Omega-3 long chain fatty acid "bioavailability": A review of evidence and methodological considerations. *Progress in Lipid Research, 56*, 92–108.

Gupta, S., Kesarla, R. and Omri, A. (2013). Formulation strategies to improve the bioavailability of poorly absorbed drugs with special emphasis on self-emulsifying systems. *ISRN Pharmaceutics*, 1–16. http://dx.doi.org/10.1155/2013/848043.

Hedge, O. J. and Bergström, C. A. (2020). Suitability of artificial membranes in lipolysis-permeation assays of oral lipid-based formulations. *Pharmaceutical Research*, *37*(6), 99–113.

Kaur, G., Molero, J. C., Weisinger, H. S. and Sinclair, A. J. (2013). Orally administered [14 C] DPA and [14 C] DHA are metabolised differently to [14 C] EPA in rats. *The British Journal of Nutrition*, *109*(3), 441–448.

Kaushik, P., Dowling, K., Barrow, C. J. and Adhikari, B. (2015). Microencapsulation of omega-3 fatty acids: A review of microencapsulation and characterization methods. *Journal of Functional Foods*, *19*, 868–881.

Khan, W., Chun-Mei, H., Khan, N., Iqbal, A., Lyu, S. W. and Shah, F. (2017). Bioengineered plants can be a useful source of omega-3 fatty acids. *BioMed Research International*, *2017*.

Khas-Erdene, Q., Wang, J. Q., Bu, D. P., Wang, L., Drackley, J. K., Liu, Q. S. and Zhou, L. Y. (2010). Responses to increasing amounts of free α-linolenic acid infused into the duodenum of lactating dairy cows. *Journal of Dairy Science*, *93*(4), 1677–1684.

Kinsella, J. E., Broughton, K. S. and Whelan, J. W. (1990). Dietary unsaturated fatty acids: Interactions and possible needs in relation to eicosanoid synthesis. *The Journal of Nutritional Biochemistry*, *1*(3), 123–141.

Köhler, A., Heinrich, J. and Von Schacky, C. (2017). Bioavailability of dietary omega-3 fatty acids added to a variety of sausages in healthy individuals. *Nutrients*, *9*(6), 629.

Köhler, A., Sarkkinen, E., Tapola, N., Niskanen, T. and Bruheim, I. (2015). Bioavailability of fatty acids from krill oil, krill meal and fish oil in healthy subjects–a randomized, single-dose, cross-over trial. *Lipids in Health and Disease*, *14*(1), 19.

Kurowska, E. M., Dresser, G. K., Deutsch, L., Vachon, D. and Khalil, W. (2003). Bioavailability of omega-3 essential fatty acids from perilla seed oil. *Prostaglandins, Leukotrienes and Essential Fatty Acids*, *68*(3), 207–212.

Lane, K., Derbyshire, E., Li, W. and Brennan, C. (2014a). Bioavailability and potential uses of vegetarian sources of omega-3 fatty acids: A review of the literature. *Critical Reviews in Food Science and Nutrition*, *54*(5), 572–579.

Lane, K. E., Li, W., Smith, C. and Derbyshire, E. (2014b). The bioavailability of an omega-3-rich algal oil is improved by nanoemulsion technology using yogurt as a food vehicle. *International Journal of Food Science and Technology*, *49*(5), 1264–1271.

Lapointe, J. F., Harvey, L., Aziz, S., Jordan, H., Hegele, R. A. and Lemieux, P. (2019). A single-dose, comparative bioavailability study of a formulation containing OM3 as phospholipid and free fatty acid to an ethyl ester formulation in the fasting and fed states. *Clinical Therapeutics*, *41*(3), 426–444.

Libinaki, R. and Gavin, P. D. (2017). Changes in bioavailability of Omega-3 (DHA) through alpha-tocopheryl phosphate mixture (TPM) after oral administration in rats. *Nutrients, 9*(9), 1042.

Loftsson, T., Ilievska, B., Asgrimsdottir, G. M., Ormarsson, O. T. and Stefansson, E. (2016). Fatty acids from marine lipids: Biological activity, formulation and stability. *Journal of Drug Delivery Science and Technology, 34*, 71–75.

Lopez-Toledano, M. A., Thorsteinsson, T., Daak, A., Maki, K. C., Johns, C., Rabinowicz, A. L. and Sancilio, F. D. (2017). A novel ω-3 acid ethyl ester formulation incorporating Advanced Lipid Technologies™ (ALT®) improves docosahexaenoic acid and eicosapentaenoic acid bioavailability compared with Lovaza®. *Clinical Therapeutics, 39*(3), 581–591.

Lopez-Toledano, M. A., Saxena, V., Legassie, J. D., Liu, H., Ghanta, A., Riseman, S. … and Sancilio, F. D. (2019). Advanced Lipid Technologies®(ALT®): A proven formulation platform to enhance the bioavailability of lipophilic compounds. *Journal of Drug Delivery*. 1957360.

Maki, K. C., Davidson, M. H., Dicklin, M. R., Ingram, K. A., Cyrowski, M., Umporowicz, D. M. and Elliott, J. G. (2003). Bioavailability of eicosapentaenoic and docosahexaenoic n-3 polyunsaturated fatty acids in salmon patties compared with capsules. *Journal of Food Science, 68*(3), 761–764.

Maki, K. C. and Dicklin, M. R. (2019). Strategies to improve bioavailability of omega-3 fatty acids from ethyl ester concentrates. *Current Opinion in Clinical Nutrition and Metabolic Care, 22*(2), 116–123.

Maki, K. C., Palacios, O. M., Buggia, M. A., Trivedi, R., Dicklin, M. R. and Maki, C. E. (2018). Effects of a self–micro-emulsifying delivery system formulation versus a standard ω-3 acid ethyl ester product on the bioavailability of eicosapentaenoic acid and docosahexaenoic acid: A study in healthy men and women in a fasted state. *Clinical Therapeutics, 40*(12), 2065–2076.

Martin, C. R. (2015). Lipids and fatty acids in the preterm infant, Part 1: Basic mechanisms of delivery, hydrolysis, and bioavailability. *NeoReviews, 16*(3), e160–e168.

Mattson, F. H., Nolen, G. A. and Webb, M. R. (1979). The absorbability by rats of various triglycerides of stearic and oleic acid and the effect of dietary calcium and magnesium. *The Journal of Nutrition, 109*(10), 1682–1687.

McClements, D. J., Decker, E. A. and Park, Y. (2008). Controlling lipid bioavailability through physicochemical and structural approaches. *Critical Reviews in Food Science and Nutrition, 49*(1), 48–67.

Morise, A., Combe, N., Boué, C., Legrand, P., Catheline, D., Delplanque, B. and Hermier, D. (2004). Dose effect of α-linolenic acid on PUFA conversion, bioavailability, and storage in the hamster. *Lipids, 39*(4), 325.

Muller, A., Schader, C., Scialabba, N. E. H., Brüggemann, J., Isensee, A., Erb, K. H. and Niggli, U. (2017). Strategies for feeding the world more sustainably with organic agriculture. *Nature Communications, 8*(1), 1290.

Nettleton, J. A. (1995). Omega-3 fatty acids and health. In: *Omega-3 Fatty Acids and Health*, Springer, Boston, MA, 64–76.

Neubronner, J., Schuchardt, J. P., Kressel, G., Merkel, M. V., von Schacky, C. and Hahn, A. (2011). Enhanced increase of omega-3 index in response to long-term n-3 fatty acid supplementation from triacylglycerides versus ethyl esters. *European Journal of Clinical Nutrition*, *65*(2), 247–254.

Nguyen, Q. V., Malau-Aduli, B. S., Cavalieri, J., Nichols, P. D. and Malau-Aduli, A. E. (2019). Enhancing omega-3 long-chain polyunsaturated fatty acid content of dairy-derived foods for human consumption. *Nutrients*, *11*(4), 743.

Offman, E., Davidson, M., Abu-Rashid, M., Chai, P. and Nilsson, C. (2017). Systemic bioavailability and dose proportionality of omega-3 administered in free fatty acid form compared with ethyl ester form: Results of a phase 1 study in healthy volunteers. *European Journal of Drug Metabolism and Pharmacokinetics*, *42*(5), 815–825.

Offman, E., Marenco, T., Ferber, S., Johnson, J., Kling, D., Curcio, D. and Davidson, M. (2013). Steady-state bioavailability of prescription omega-3 on a low-fat diet is significantly improved with a free fatty acid formulation compared with an ethyl ester formulation: The Eclipse II study. *Vascular Health and Risk Management*, *9*, 563.

Ottestad, I., Nordvi, B., Vogt, G., Holck, M., Halvorsen, B., Brønner, K. W. and Ulven, S. M. (2016). Bioavailability of n-3 fatty acids from n-3-enriched foods and fish oil with different oxidative quality in healthy human subjects: A randomised single-meal cross-over study. *Journal of Nutritional Science*, *5*(43), 1–8.

Parada, J. and Aguilera, J. M. (2007). Food microstructure affects the bioavailability of several nutrients. *Journal of Food Science*, *72*(2), R21–R32.

Patenaude, A., Rodriguez-Leyva, D., Edel, A. L., Dibrov, E., Dupasquier, C. M. C., Austria, J. A. and Pierce, G. N. (2009). Bioavailability of α-linolenic acid from flaxseed diets as a function of the age of the subject. *European Journal of Clinical Nutrition*, *63*(9), 1123–1129.

Poumès-Ballihaut, C., Langelier, B., Houlier, F., Alessandri, J. M., Durand, G., Latge, C. and Guesnet, P. (2001). Comparative bioavailability of dietary α-linolenic and docosahexaenoic acids in the growing rat. *Lipids*, *36*(8), 793–800.

Punia, S., Sandhu, K. S., Siroha, A. K. and Dhull, S. B. (2019). Omega 3-metabolism, Absorption, bioavailability and health benefits-A review. *PharmaNutrition*, *10*, 100162.

Qin, Y., Nyheim, H., Haram, E. M., Moritz, J. M. and Hustvedt, S. O. (2017). A novel self-micro-emulsifying delivery system (SMEDS) formulation significantly improves the fasting absorption of EPA and DHA from a single dose of an omega-3 ethyl ester concentrate. *Lipids in Health and Disease*, *16*(1), 204.

Rein, M. J., Renouf, M., Cruz-Hernandez, C., Actis-Goretta, L., Thakkar, S. K. and da Silva Pinto, M. (2013). Bioavailability of bioactive food compounds: A challenging journey to bioefficacy. *British Journal of Clinical Pharmacology*, *75*(3), 588–602.

Sanguansri, L., Shen, Z., Weerakkody, R., Barnes, M., Lockett, T. and Augustin, M. A. (2013). Omega-3 fatty acids in ileal effluent after consuming different foods containing microencapsulated fish oil powder–an ileostomy study. *Food and Function*, *4*(1), 74–82.

Schuchardt, J. P. and Hahn, A. (2013). Bioavailability of long-chain omega-3 fatty acids. *Prostaglandins, Leukotrienes and Essential Fatty Acids*, 89(1), 1–8.

Schuchardt, J. P., Schneider, I., Meyer, H., Neubronner, J., von Schacky, C. and Hahn, A. (2011). Incorporation of EPA and DHA into plasma phospholipids in response to different omega-3 fatty acid formulations-a comparative bioavailability study of fish oil vs. krill oil. *Lipids in Health and Disease*, *10*(1), 145.

Staprans, I., Rapp, J. H., Pan, X. M., Kim, K. Y. and Feingold, K. R. (1994). Oxidized lipids in the diet are a source of oxidized lipid in chylomicrons of human serum. *Arteriosclerosis and Thrombosis: A Journal of Vascular Biology*, *14*(12), 1900–1905.

Stumpf, P. K. (1969). Metabolism of fatty acids. *Annual Review of Biochemistry*, *38*(1), 159–212.

Taha, A. Y., Cheon, Y., Faurot, K. F., MacIntosh, B., Majchrzak-Hong, S. F., Mann, J. D. and Ramsden, C. E. (2014). Dietary omega-6 fatty acid lowering increases bioavailability of omega-3 polyunsaturated fatty acids in human plasma lipid pools. *Prostaglandins, Leukotrienes and Essential Fatty Acids*, *90*(5), 151–157.

Ulven, S. M. and Holven, K. B. (2015). Comparison of bioavailability of krill oil versus fish oil and health effect. *Vascular Health and Risk Management*, *11*, 511.

Wakil, A., Mackenzie, G., Diego-Taboada, A., Bell, J. G. and Atkin, S. L. (2010a). Enhanced bioavailability of eicosapentaenoic acid from fish oil after encapsulation within plant spore exines as microcapsules. *Lipids*, *45*(7), 645–649.

Wakil, A., Mir, M., Mellor, D. D., Mellor, S. F. and Atkin, S. L. (2010b). The bioavailability of eicosapentaenoic acid from reconstituted triglyceride fish oil is higher than that obtained from the triglyceride and monoglyceride forms. *Asia Pacific Journal of Clinical Nutrition*, *19*(4), 499.

Walker, R., Decker, E. A. and McClements, D. J. (2015). Development of food-grade nanoemulsions and emulsions for delivery of omega-3 fatty acids: Opportunities and obstacles in the food industry. *Food and Function*, *6*(1), 41–54.

Wallace, J. M. W., McCabe, A. J., Robson, P. J., Keogh, M. K., Murray, C. A., Kelly, P. M. and Strain, J. J. (2000). Bioavailability of n–3 polyunsaturated fatty acids (PUFA) in foods enriched with microencapsulated fish oil. *Annals of Nutrition and Metabolism*, *44*(4), 157–162.

Wood, K. E., Lau, A., Mantzioris, E., Gibson, R. A., Ramsden, C. E. and Muhlhausler, B. S. (2014). A low omega-6 polyunsaturated fatty acid (n-6 PUFA) diet increases omega-3 (n-3) long chain PUFA status in plasma phospholipids in humans. *Prostaglandins, Leukotrienes and Essential Fatty Acids*, *90*(4), 133–138.

Yang, Q., Wang, S., Ji, Y., Chen, H., Zhang, H., Chen, W. and Chen, Y. Q. (2017). Dietary intake of n-3 PUFAs modifies the absorption, distribution and bioavailability of fatty acids in the mouse gastrointestinal tract. *Lipids in Health and Disease*, *16*(1), 10.

Yehuda, S., Rabinovitz, S. and Mostofsky, D. I. (2005). Essential fatty acids and the brain: From infancy to aging. *Neurobiology of Aging*, *26*(1), 98–102.

Yep, Y. L., Li, D., Mann, N. J., Bode, O. and Sinclair, A. J. (2002). Bread enriched with microencapsulated tuna oil increases plasma docosahexaenoic acid and total omega-3 fatty acids in humans. *Asia Pacific Journal of Clinical Nutrition*, *11*(4), 285–291.

9

Enrichment of Essential Fatty Acids in Food

Manju Nehra, Kiran Bala Nain, Anju Malik, Mohd. Mohsin and Amanjyoti Gill

Contents

9.1 INTRODUCTION

Many people consider that using fat in lesser quantities is better for health. This concept may true in the case of saturated fat but does not fit for the consumption of a diet rich in unsaturated fatty acids (Hennalu and Norziah 2011). Polyunsaturated fatty acids (PUFAs), which play very vital roles in human health from conception through pregnancy, childhood and the whole lifespan, are not synthesized within the body but are supplied through diet, and are referred to as essential fatty acids (EFAs) (Ganesh and Hettiarachchy 2016). The health benefits of essential fatty acids to human beings are universally recognized, and awareness among the people for their wellbeing is the main cause of the growing trend of EFA-enriched food products in the global market during recent years. As these are not synthesized within the human body, most of the requirements are met by diet. A major part of the dietary supply is associated with animal sources, especially fish. This is problematic for the people who do not like to eat fish. Various deficiency diseases are reported in the literature that are due to a paucity of essential nutrients, and this enforces the enrichment of food with required nutrients for confirming the sound health of consumers. Many establishments have made certain recommendations for society to enhance its intake of PUFAs; therefore, due to advancements in technology allowing producers to securely enrich/fortify food items with long-chain polyunsaturated fatty acids, these are added into various non-traditional food items on a large scale (Whelan and Rust 2006). Therefore, food enrichment/supplementation with essential fatty acids seems to be the most effective path to meet the recommended amount. Recently, trends have been focused toward the incorporation of fatty acids into popular foods. The latest concept in the enrichment of foods with EFAs has been achieving popularity among food scientists, technologists, researchers and health-conscious consumers owing to the wonderful health benefits of direct ingestion of EFA-supplemented foods.

Enrichment is the process of incorporation/addition of nutrients previously present in foods but lost during processing. Supplementation of generally eaten food items with PUFAs is thought to be an innovative approach for offering health advantages to people without requiring major variation in their dietary behaviors (Garg et al. 2006; Anbudhasan et al. 2014). Food enrichment with marine omega-3 PUFAs presents a multidisciplinary challenge for academia and the food industry (Jacobsen 2010a) because of the highly oxidative nature of unsaturated fatty acids. Fatty acid–rich oils, either directly added using many processing techniques or biodelivered by feeding animals with fatty acid–rich products, may be an alternative for the enrichment of food products (Whelan and Rust 2006). Technical advancements like the development of microencapsulation, nanoencapsulation and new antioxidants to protect the taste and flavor of PUFA-enriched foods are the major cause for their growing market potential. In this chapter, we will emphasize certain processing techniques which are helpful in the alteration of the EFA composition of different food items in such a manner that they become more available to the consumer.

9.2 NEED OF FOOD ENRICHMENT WITH EFAS

EFAs are naturally found in many food products, but during processing and preservation some of these become depleted or lost, which alters the nutritional composition of food and may lead to various deficiency diseases such as impairment of mental and visual health, nervous and immune system disorders, as well as cardiovascular and skin-related problems, etc. (Jacobsen 2010b). Recently, to combat the deficiency of EFAs in raw food or in powdered products, the enrichment of foods either by modifying the diet of animals or by use of external sources has been carried out on a large scale.

From a nutritional point of view, n-3 (omega-3) and n-6 (omega-6) are two major kinds of naturally occurring EFAs. Among the three most important n-3 fatty acids, i.e., alpha-linolenic acid (ALA, C18:3), eicosapentaenoic acid (EPA, C20:5) and docosahexaenoic acid (DHA, C22:6), EPA and DHA are well reported for their multiple health-promoting properties (Fraeye et al. 2012). Fish, shellfish, fatty fish, sea mammals, algae and fish oil are the most popular and natural sources of n-3 EFAs (especially EPA and DHA). Moreover, vegetable oils like canola, flax seed, safflower, sunflower, sesame, corn, palmolein and unseen fat in pulses, legumes, cereals, tubers and vegetables provide good amounts of n-6 fatty acids (Mantzioris et al. 2000). ALA can be obtained mainly from seeds and oils, green leafy vegetables, walnuts and soybeans (Ganesan et al. 2014). Vegetable oils are rich sources of ALA but are deficient in EPA and DHA. Owing to inadequate ingestion of fish by non-vegetarians and more amounts of n-6 fatty acids in foods generated by recent intensive production approaches, the required quantity of n-3 fatty acids cannot be fulfilled by dietary sources, which creates a huge imbalance in the ratio of n-6:n-3 fatty acids in the modern-day diet (Cherian 2013).

In the year 2000, the Food and Drug Administration recommended that consumption of EPA and DHA from food and dietary supplements, fish oil and other sources should not go above 3.0 g/day for a person, but recently, in countries having a Western-style diet, n-3 PUFAs are nearly about 0.15 g/day, which is far less than the suggested amount (Nettleton 1995; FDA 2000; Sanders 2000; Kris-Etherton et al. 2000; Kolanowski and Laufenberg 2006). To meet the suggested dietary requirements of PUFAs and compensate for the processing losses, there is a great demand for developing n-3-enriched foods for a balanced n-3/n-6 diet. Levels of n-3 PUFAs recommended by different organizations are represented in Table 9.1.

At present, in the United States, Europe, India and other countries of the world, functional foods having n-3 lipids comprise one of the fastest-increasing food product classes; many of these foods are already available

Table 9.1 Recommended Levels of Essential Fatty Acids by Different Organizations

S. No.	Age Groups	EPA + DHA/Day (mg)
FAO (United Nations) Guidelines		
1	Adults	250 mg
2	Pregnant/lactating women	300 mg, 200 mg out which should be DHA
3	Children 2–4 years old	100–150 mg
4	Children 4–6 years old	150–200 mg
5	Children 6–10 years old	200–250 mg
International Association for the Study of Fatty Acids and Lipids		
1	Adults	500 mg/day EPA or DHA (or combination)
American Heart Association		
1	General public	Two servings of fatty fish per week
2	Cardiovascular patients	1000 mg/day of EPA or DHA (or combination)
3	Patients with high triglycerides	2000–4000 mg/day of EPA or DHA (or combination)

(Compiled from Simopoulos et al. 2000; Kris-Etherton et al. 2003; Gouin 2004; Feizollahi et al. 2018)

in the market (Jacobsen 2010b). Usage of fish oils in food products is a rapidly developing industry owing to increasing evidence about the health-boosting potential of EPA and DHA. Additionally, seafood consumption by people at many places is negligible, resulting in too little ingestion of essential fatty acids through the diet (Jacobsen 2010a). Also, decreasing fish stocks, fishy smell, seasonal changes in quality and fish contamination with heavy metals and polychlorinated biphenyls (PCBs) are some other reasons responsible for reduced consumption of fish (Belarbi et al. 2000; Qi et al. 2004; Domingo and Bocio 2007; Babuskin et al. 2014). Thus, there is a great need for the enrichment of food items with EFAs to meet the dietary requirements. Various emerging techniques like molecular complexing, genetic engineering, encapsulation and nanotechnology have recently been employed for the enrichment of foods with EFAs.

9.3 ENRICHMENT OF FOODS WITH N-3 PUFAS

Raw foods are subjected to hundreds of processing treatments, adopted worldwide to make foods more palatable and available for consumption of the people. Basic processing techniques may also be responsible for the alteration of nutritional composition of food products which results in nutrient deficiency. Common processed foods like cakes, ice cream, bread, biscuits, milkshakes, orange juice, spreads, mayonnaise, pasta and vinaigrette and so on can be enriched with n-3 PUFAs using various techniques. These advanced methods do not change the final taste of the foods and provide health benefits to the consumers.

Presently, most of the enriched foods found in the market are emulsion based such as milk and other dairy items, dressings, spreads and mayonnaise, etc. (Jacobsen et al. 2013a). Distribution of n-3 PUFAs in foodstuffs can be enhanced by following two potential approaches, that is, providing n-3-enriched diets to the animals for intensifying the natural content of essential fatty acids in meat, eggs, fish, milk and so on, or preparation of processed food items by incorporating ingredients rich in n-3 PUFAs (Ansorena and Astiasaran 2013). Processed food products can be enriched with omega-3 PUFAs using microencapsulated fish oil, neat oils, or a fish oil-in-water emulsion, as different food items are compatible with different forms. Milk, yogurt and dressing treated with emulsified fish oil and neat oil resulted in products with different oxidative stabilities, which were presented as classic examples of fortified foods (Jacobsen 2010a). Examples of some fish oil–enriched food products are shown in Table 9.2.

Table 9.2 Examples of Fish Oil-Enriched Foods

Product Type	Examples
Bread and bakery products	Nimble Heartbeat (British Bakeries, UK) Nutribread (William Jackson Bakery, UK) Omega bread, sponge cakes, biscuits (Functional Nutrition, UK) Omega-3-bread and omega-3-rolls (VK Muhlen, Germany) Diamant Vital omega-3 Kruste (Diamant Mühle, Germany) Wellness Aktifit-Brot (RUF Lebensmittelwerke, Germany) Leva omega-3-bread (Pagen, Sweden) F Plus fortified biscuits (Cuetara, Spain) Omega bread (Piekarnia Parkowa Wroclaw, Poland), TipTop Up sliced bread and muffins (TipTop bakeries, Australia) Hi Q DHA bread (Bunge Defiance, Australia), DHA-bread and DHA table rolls (Yamazaki Baking, Japan)
Milk and dairy products	Omega-3 milk (Parmalat, USA) Lactel omega-3 milk (LactalisBesnierBridel, France) Mleko omega-3 – UHT milk (SM Sudowa, Poland) Laukiomega+skim milk (Candia, Spain) Especial omega-3 milk (Mimosa, Portugal) Plus omega-3-latte and omega-3-yogurt (Parmalat, Italy) Omi-3 yogurt (SM Siedlce, Poland) Omi-3 processed quark (OSM Ostrołka, Poland)
Spreadable fats	Gaio Spread (MD foods, Scandinavia) Omega fat spread (Fjordland, Norway) Vitaquell omega-3 (Vitaquell, UK) Heartwatch omega reduced fat spread (Functional Nutrition, UK) Vigor Omega Vitta (Vigor, Brazil) Dos Alamos Margarina (Grasco, Chile)
Eggs and egg products	DHA Gold eggs (OmegaTech, USA) Gold Circle Farms eggs (NutraSweet Kelco, USA) Eggs Plus (Pilgrim's Pride, USA) Ovo3 eggs (Maia Agrolimentare, Italy) Oro omega-3 eggs (Unione Cascine Valpadana, Italy) Columbus eggs (Belovo, Netherlands) Eiplus eggs (Eifrisch, Germany) Minicol omega pasteurized eggs (Wammala Food, Finland)

(*Continued*)

Table 9.2 (Continued) Examples of Fish Oil-Enriched Foods

Product Type	Examples
Meat and poultry products	Strasburg sausage (Hans, Australia) Omega-3 Jamon Cocido cooked ham breast (Carnicas Serrano, Spain) Terra y Mar turkey breast (Carnicas Serrano, Spain) Omega cool burger (Pals, Norway) Mega Off frozen chicken pieces enriched (Off Tene, Israel)
Juices and soft drinks	Tidal Wave Superfood Juice (Naked Juice, CA, USA) Supajus DHA-rich orange drink (The Natural Fruit & Beverage Co., UK) My way wellness drink (Designer Foods, Germany) Timlic – orange nectar (Bauman-Fruits, Switzerland) ChikaraMizu soft drinks (Kirin Breweries, Japan)

(Adapted from Kolanowski and Laufenberg 2006)

There are many methods available for the extraction of EFAs from foods using solvents like petroleum benzene, methanol-water-petroleum benzene, supercritical fluids such as carbon dioxide, hexane and so on. These extracts are further incorporated into convenient foods for fulfilling the dietary needs of the body as indicated by a literature survey. For the extraction of PUFAs (especially EPA and DHA) from freshwater and marine fish in pure form, supercritical fluid extraction is a promising and currently the best technique being used. Generally, PUFAs are extracted at about 10–30 MPa and 40–80°C (Sahena et al. 2009).

The choice of mode of delivery for n-3 PUFAs plays an important role and therefore must be considered very carefully. Some modern techniques used for increasing the amount of n-3 PUFAs in foods include enzymatic splitting, low-temperature crystallization, fractional and molecular distillation, supercritical fluid extraction, chromatographic techniques and urea complexation. Omega-3 PUFA concentrates can be found in various forms, that is, free fatty acids, methyl and ethyl esters or acylglycerols (Shahidi and Wanasundra 1998; Kolanowski and Laufenberg 2006). For obtaining oil concentrates rich in PUFAs, low-temperature treatment (winterization) is commonly used. Further purifications can be continued using enzymatic hydrolysis. For separation of unsaturated fatty acids from mixture urea complexation, liquid chromatography and enzyme-based process are used. Selective esterification of free fatty acids (PUFA transformations) is another way of oil purification. Table 9.3 represents some general PUFA enrichment processes used for obtaining purified extracts.

9.3.1 Enrichment of Foods with n-3 PUFAs by Modifications in Animal Diet

Animals and humans beings cannot efficiently convert $C_{18:2n-6}$ and $C_{18:3n-3}$ into n-3 PUFAs, mainly $C_{20:4n-6}$, EPA and DHA, so from the diet these get accumulated in the edible body parts. The influence of animal feeding on human foods is much more than the effect of vegetable-growing process on the nutritional composition of foods. The fatty acid profile of any animal depends on its species and breed, but even more on the diet (Bourre, 2005). Thus, it is more significant to enrich animal feed with n-3 PUFA-rich diets having linseed, linseed oil, rapeseed, fish products, etc. On the addition of EPA and DHA to a non-ruminant's diet, these are quickly absorbed and accumulated in animal products. Animal products may

Table 9.3 Summary of General PUFA Enrichment Processes

Method	Procedure
Winterization	Reduce temperature to render more saturated fats insoluble, separate out insoluble fraction
Glyceride hydrolysis	Hydrolyze glycerides to glycerol and free fatty acids using saponification or lipase enzymes
Urea complexation	Add urea and aqueous ethanol, heat to solubilize the fatty acid mixture, reduce temperature to crystallize urea-saturated complexes, centrifuge/filter out crystals, acidify supernatant with hexane, recover PUFAs as free fatty acids from hexane by evaporation
Enzyme splitting	Promote lipase selective esterification of more saturated fatty acids, separate concentration of PUFAs in free fatty acid fraction from esterified saturated fatty acids, repeat esterification for further PUFA enrichment
PUFA transformations	Esterify PUFAs to produce esters, interesterifications to enrich non-highly unsaturated fatty acid oils with PUFAs

(Adapted from Beligon et al. 2016)

still be improved and considerably contribute to the total dietary intake of essential fatty acids (Jakobsen 1999). Linseed (*Linum usitatissimum*), which contains nearly 40% oil, is one of the chief oilseed plants; plus, fish oil, microalgae and marine algae are commonly used as a source of omega 3 fatty acids for animal feeding (Woods and Fearon 2009; Givens et al. 2000; Dewhurst and Moloney 2013). Bio hydrogenation, which is responsible for the conversion of dietary fatty acids into saturated fats, was less extensive in sheep as compared to cattle, and dietary manipulation was the most effective way for changing the fatty acid composition of sheep meat also (Chikwanha et al. 2017). For sound consumer health, processes of enrichment of meat and milk with fatty acids using animal diet manipulations no doubt may affect the health and fertility of animals. However, there are also some issues related to cost of n-3 PUFA (linseed and fish oil)-rich animal feed and processing of feeds to decrease rumen bio-hydrogenation. Thus, plants provide the lowest-cost approach, and future research must focus on the reduction of rumen bio-hydrogenation of herbage PUFAs and enhancement of enzyme activities used in n-3 PUFA synthesis in the mammary gland and muscles (Dewhurst and Moloney 2013).

Commercial table eggs are found rich in n-6 PUFA (mainly 18:2n-6), but these eggs contain much lower quantities of n-3 fatty acids. Therefore, using various techniques, these are fortified with essential fatty acids (Surai and Sparks 2001). Flaxseed (rich in α-linolenic acid) was traditionally used for feeding hens, and recently fish oil (rich in EPA and DHA), as well as microalgae (rich in EPA or DHA) are used for egg enrichment purpose (Fraeye et al. 2012). For achieving the recommended n-6/n-3 ratio of 3–5:1, n-3 PUFA-enriched eggs were produced by feeding laying hens with varying quantities of linseed oil (1%, 2%, 3% or 4%) in the diet which resulted in a decreased ratio during the first 5 weeks and stability after that (Petrovic et al. 2012). The concentration of ALA become enhanced nearly 2-fold in beef, 6-fold in pork, 10-fold in chicken and 40-fold in eggs when animals were fed with extracts of linseed and rapeseed grains. Furthermore, the quantity of DHA was raised approximately 2-fold in case of beef, 7-fold in chicken, 6-fold in eggs and 20-fold in salmon fish when fish extracts or algae oils were given to the animals (Bourre 2005). Eating only one n-3-enriched egg on a daily basis can significantly influence the recommended intake of n-3 PUFAs, as the ratio of n-6 to n-3 in plasma was decreased significantly from 12.2:1 to 6.5–7.7:1 using n-3-enriched eggs produced by feeding hens with fish oil or a combination of fish and vegetable oil. However, the sensory quality and shelf life of enriched eggs were not affected during storage (Farrell 1998). White Leghorns (laying hens) were fed with diets containing 10% fish, linseed or safflower oil and controls with 2.9% of unidentified fats. The effect of adding different oils to the diet on the composition of the eggs laid is represented in Table 9.4.

Diet, whether natural or compounded, is the major factor associated with the fatty acid composition of fish, but species, size, genetics, season, location and environment also affect the lipid profile (Justi et al. 2003). In addition, marine fish farming is a very helpful way of providing fish rich in n-3 PUFAs without any health hazards related to fish consumption, as fish are reared and collected under control conditions. Furthermore, the correct selection of lipid sources for fish feeding during culturing is very valuable for tailoring the foods according to consumer demand (Alasalvar et al. 2002; Fuentes et al. 2010; Hossain 2011). Sunflower, linseed and linseed and olive (1/1, wt/wt) were used as oil sources in animal feed, and with linseed-enriched batches, a good ratio of PUFA n-6/n-3 (< 4) was achieved; furthermore, pork liver paté with back fat and liver enriched with n-3 PUFAs and α-tocopherol were prepared (Arrigo et al. 2004). Currently, from the fat point of view, red meat is considered unhealthy,

Table 9.4 Effect of Adding Oils to the Diet of Laying Hens on Egg Composition

		Laying Hens			
Fatty Acids	**Control**	**Fed Fish Oil**	**Fed Linseed Oil**	**Fed Soybean Oil**	**Fed Safflower Oil**
ALA					
% total fatty acids	0.4	0.4	9.4	1.7	0.06
EPA					
% total FA	–	2.1	0.3	–	–
DHA					
% total FA	0.6	6.3	2.2	1.8	0.2
Total ω-3	0.6	10.1	12.6	3.1	0.3
% FA					
Total ω-6 + ω-3					
% FA	18.8	19.7	32.3	28.3	36.2
ω-6/ω-3	17.8	1.0	1.6	8.1	120

(Adapted from Bourre 2005)

as the saturated fatty acid level of meat fat is very high and PUFA content is low. Making use of late-maturing breeds and killing them at lighter weights is a good approach to enhance the n-3 PUFA content in meat animals. However, economically it may not be desirable in numerous manufacturing environments (Dewhurst and Moloney 2013). By changing the feeding habits (providing n-3 PUFA-rich diets) of cattle, the fatty acid pro file can be altered effectively.

Processed tuna fishmeal might be a good alternative source of DHA for enrichment of pork, chicken and eggs with n-3 PUFAs. The use of tuna fishmeal (10% PorcOmega™, a by-product of the tuna processing industry) in rations of laying hens, broilers and pigs caused increased levels of fatty acids (up to 7-fold) in chops, sausages and eggs. However, the use of higher levels (exceeding 15% in meat and 10% in eggs) might cause some sensory issues (Howe et al. 2002). Saturated and monounsaturated FAs were decreased, whereas PUFA contents (mainly linolenic and linoleic) were increased with enhanced levels of linseed oil in the diet of chicken meat and even after feeding linseed oil for a long time, peripheral tissue deposition of n-3 PUFAs was continued (Lopez-Ferrer et al. 2001). Egg yolk color of laying hens was strongly influenced by microalgal (*Isochrysis*) feeding, owing to the transfer of carotenoids from the microalgal biomass, and a dose of 120 mg n-3 PUFAs/100 g is optimum, which is obtained from 2.4% of *Isochrysis* biomass for maintaining acceptable yolk color (Lemahieu et al. 2014).

9.3.2 Enrichment of Foods with n-3 PUFAs Using Microencapsulation

Recently, microencapsulation has been considered to be a useful technique to slow down the fat oxidation of PUFAs, mask the undesirable taste of some ingredients, attenuate the production of fishy flavors, and raise the shelf life of foods, as well as to make the fortification process better for the foods. This technique protects the core compound by allowing a bioactive compound to be entrapped or encapsulated within a minute sphere called a microsphere or microcapsule (Anbudhasan et al. 2014). Spray-drying, freeze-drying, spray-cooling, liposome entrapment, complex coacervation, supercritical-fluid-based techniques, fluidized-bed coating and extrusion-based techniques are commonly employed microencapsulation techniques utilized for n-3 PUFA oil encapsulation (Feizollahi et al. 2018). The prevention of oxidation is the main point which must be considered during enrichment. The microencapsulated products might be used in emulsion or powder form. In the case of emulsions, the n-3 PUFA oil-containing phase was found to be dispersed in other continuous phases with an emulsifier.

Fish oil is the oil achieved from the flesh of oily fishes or from the liver of lean fishes, which comprise EPA and DHA, and 1 g fish oil capsule contains EPA and DHA as nearly 30% of the fatty acids found; moreover, an EPA:DHA ratio of 1.5:1 is found in most standard fish oils (Calder 2013). Fish oil may be used directly or can be used in combination with more saturated oils like canola, high-oleic sunflower, olive, palm and partially hydrogenated oils. Metal contact should be negligible. Various types of encapsulated systems used for oil encapsulation are shown in Table 9.5.

Preparation of n-3 PUFA (mainly EPA and DHA)-enriched spaghetti with high nutritional value and sensory acceptability using microencapsulated integrator (intermediate levels of integrator i.e. 1.2%) might fulfill 20% of the daily recommended intake of fatty acids along with minimum cooking losses (Iafelice et al. 2008). The palatability of food items by fish oil enrichment was affected very differently, and an objectionable fishy taste could be masked with natural strong flavor and sweetness of foods as well as low-pH liquid foods like juices, which require more advanced techniques for oxidative stability. Instant powder-form formula concentrates can be enriched suitably with microencapsulated fish oil with a 6-month storage stability (Kolanowksi et al. 1999). The influence of substitution of shortening with microencapsulated n-3 PUFA powder (1%) in whole wheat breads indicated a high (80–89%) EPA and DHA recovery and less fat oxidation after baking; during storage, the stability of microencapsulated n-3 PUFA powder in bread was confirmed (Hennalu and Norziah 2011). The literature also confirmed that bread enriched with microencapsulated tuna oil (as a delivery vehicle) enhances the plasma DHA and total n-3 fatty acids in humans with 60 mg/day bioavailability (Yep et al. 2002). The bioavailability of n-3 PUFAs in foods enriched with microencapsulated fish oil was compared with fish oil capsules, and results confirmed that the bioavailability of n-3 PUFA from microencapsulated fish oil–enriched foods was similar to that of fish oil capsules (Wallace et al. 2000). Therefore, microencapsulation offers an effective route for enhancing n-3 PUFA consumption.

Table 9.5 Several Encapsulation Systems Used to Encapsulate n-3 Fatty Acids

Encapsulation System	Encapsulated Ingredient	Shell Material	Product Properties	References
Single-layered o/w emulsion	Algal oil	Whey protein	Utilized as a direct additive to yogurt, consumer panel could not distinguish strong fish flavor after 22 days storage	Chee et al. (2005)
Single-layered o/w emulsion	Tuna oil	Chitosan, maltodextrin, whey protein	Droplet size around 10 µm after 30 days storage, encapsulation efficiency >80%, low water activity	Klaypradit and Huang (2008)
Spray-dried single-layered o/w emulsion	Fish oil	Pectin	Droplet size <2 µm, oxidative stability strengthened	Drusch (2007)
Spray-dried single-layered o/w emulsion	Fish oil	Barley protein	Encapsulation efficiency >92.9%, loading efficiency 46.5–50.1%, omega-3 oils stable at 40°C for 8 weeks	Wang et al. (2011)
Freeze-dried single-layered o/w emulsion	Sandeel oil	Sodium caseinate, lactose, maltodextrin	Particle size <1 µm, more oxidative stable	Heinzelmann et al. (2000)
Spray-dried multilayered o/w emulsion	Tuna oil	Lecithin, chitosan	Droplet size 5–30 µm, low moisture, oil retention levels >85%, water dispensability <1 min, strengthened oxidative stability	Klinkesornet al. (2006)
w/o/w multiple emulsion	Fish oil	Insulin, Arlatone F 127	Excellent oxidative stability during storage at 4°C for 6 weeks	Cournarie et al. (2004)

(Adapted from Barrow and Wang 2013)

9.3.3 Enrichment of Processed Foods Using Plant or Microalgal Sources

The microorganisms (fungi, yeast, microalgae and thraustochytrids) provide a promising way of manufacturing PUFAs with higher yield and lower cost. Thus, choosing microflora efficiently producing fatty acids, selecting proper metabolic pathways and using low-cost substrates, single-cell oils can be prepared and used as dietary supplements (Beligon et al. 2016). Certain algal oils are mainly abundant in DHA, which may contain as much as 45% of fatty acids available, and these oils are used in infant formulas, in which DHA is especially desired (Calder 2013). There are many algal species which contain high amounts of DHA; however, the high price of oil extraction from krill and algae and costly cultivation as well as harvesting are the main reasons for the limited use of algae as a source of n-3 PUFAs for enrichment purposes (Feizollahi et al. 2018). Functional cookies and pasta can be prepared by incorporating freeze-dried *Nannochloropsis oculata* (marine microalga) biomass as a source of n-3 fatty acids (EPA + DHA) owing to its high growth rate, lipid profile, nutritional benefits and resistance to mixing and contamination. Cookies supplemented with 2% of microalgal biomass (197 mg/100 g PUFAs) and pasta with 3% biomass (190 mg/100 g PUFAs) might be a good option for enrichment purposes, along with providing a 2-month storage stability of color and n-3 fatty acid profiles (Babuskin et al. 2014).

DHA and arachidonic acid (AA) are important constituents of breast milk and are advised for supplementation of commercial infant formula, as these are safe and promote growth and are very effectively used in the development of visual and neurological systems (Kuratko et al. 2013). Cow's milk is deficient in DHA and AA; thus, enrichment with these fatty acids is necessary during the preparation of infant formulas. Owing to unsaturation in DHA and AA structures, they are very vulnerable to peroxidation and reactions with the other nutrients. Consequently, some special methods are used for their incorporation in infant formula. A valuable and reproducible technique, i.e., the use of nano-liposomes for the application of n-3 PUFAs in the food system, was developed (Rasti et al. 2017). Milk and bread can be enriched with EPA and DHA in the form of nano-liposomes. From the sensory point of view, nano-liposomal n-3 PUFA-containing milk and breads were similar to the control samples and showed high storage stability without any detectable foul fishy smell.

9.4 ROLE OF ENZYMES IN ENRICHMENT OF FOOD WITH N-3 PUFAS

A potential enzyme, PLA1, can be used for selective concentration of n-3 PUFAs. For preparing n-3 PUFA-enriched supplements, lipases are used which liberate n-3 fatty acids from the oils. However, the use of lipases is limited owing to insufficient specificity for n-3 PUFAs (Moharana et al. 2016). In canola (*Brassica napus*) seeds, a microalgal polyketide synthase-like PUFA synthase system, including an accessory enzyme and three multidomain polypeptides, was expressed by Walsh et al. (2016). This enzymatic system synthesized DHA and EPA de novo from malonyl-CoA without greatly changing plastidial fatty acid preparation in the cytoplasm. It was also reported that canola oil processed from field-grown grain contains DHA (3.7%) and EPA (0.7%) which can give greater than 600 mg of n-3 PUFAs in a 14-g serving. Approaches for the enzymatic enrichment of PUFAs from fish oil (Nile perch oil) by Mbatia et al. (2011) suggested that for producing the products rich in DHA and EPA, hydrolysis of simple esters and esterification of free fatty acids (lipase-catalyzed methods) were very efficient, and loss of essential fatty acids during the enzymatic processes could be easily compensated by side streams and recycled for highest recovery.

9.5 ROLE OF BIOTECHNOLOGY IN PRODUCTION OF PUFA-RICH FOOD PRODUCTS

Marine bacteria, oleaginous fungi, heterotrophic and phototrophic microalgae and mosses are rich in PUFAs and can be used as an alternative to agricultural and animal sources. Thus, the newly prepared fatty acid–rich foods using solid-state fermentation (SSF) and transgenic cereals using gene engineering may open innovative market possibilities in the future. The use of biotechnological techniques based on SSF and genetic engineering for the preparation of functional cereals enriched with PUFAs and carotenoids in a natural way is shown in Figure 9.1 (Certik et al. 2013a).

For enrichment of cereals with n-3 PUFAs and pigments using fungal solid-state fermentations, four *Mucor* strains were analyzed for their capacity to grow on four cereal substrates and were enriched with gamma-linolenic acid (GLA). The maximum quantity of GLA was obtained by the use of 30% sunflower oil in the fermented substrate (Certik et al. 2013b). The ability of fungus *Thamnidium elegans* and yeast *Rhodotorula glutinis*

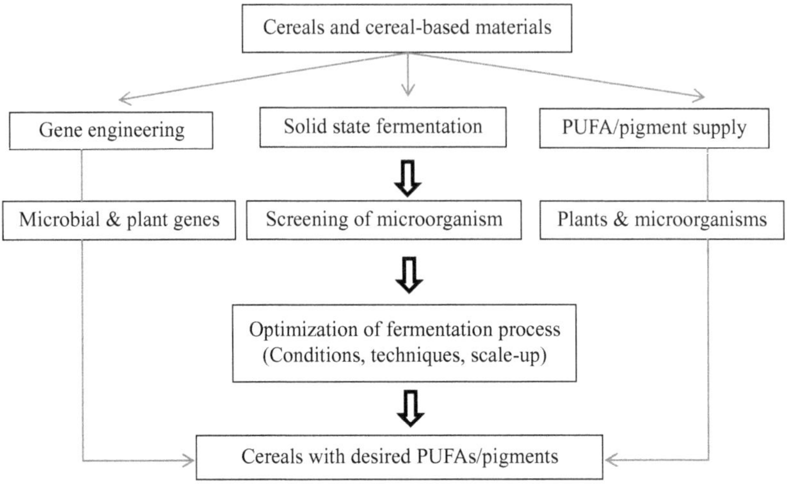

Figure 9.1 Biotechnological approaches for modification of cereals enriched with PUFAs and pigments (adapted from Certik et al. 2013b)

and *Sporobolomyces roseus* to use cereal substrates in solid-state fermentation was also tested (Certik et al. 2008), which demonstrated that cereal products could be enriched with PUFAs very effectively and newly prepared bio-products could be used for feed, food, veterinary and pharmaceutical fields.

9.6 BIOAVAILABILITY OF ESSENTIAL FATTY ACIDS FROM N-3-ENRICHED FOODS

The physiological influence and bioavailability of foods enriched with n-3 PUFAs depend on the type and amount of other nutrients found in foods. The contribution of dietary ALA and linoleic acid (LA) in long-chain PUFA biosynthesis and tissue composition of rats after feeding for 8 weeks was assessed and ALA-rich diet was reported with more n-3 PUFAs including DHA in brain and heart, whereas the amount of n-3 in liver, heart and blood was decreased with LA diet, although no change in n-6 PUFA composition was recorded (Blanchard et al. 2013). Thus, by decreasing LA intake and increasing dietary consumption of ALA, the bioavailability of n-3 PUFAs can be enhanced. Omega-3 fatty acids from microencapsulated fish oil powder and ready-to-eat meals enriched with liquid fish oil possessed similar bioavailability (Hinriksdottir et al. 2014), which indicated a beneficial use of encapsulated powder for enhancing the n-3 PUFA concentration in blood. The availability of n-3 PUFA-enriched processed foods could substantially enhance the essential fatty acid levels of consumers. The replacement of food with n-3 PUFA-enriched foods for 6 months by overweight people who were low fish eaters causes a greater increase in total PUFAs, EPA and DHA (Patch et al. 2005).

9.7 INFLUENCE OF N-3 PUFA-ENRICHED DIETS ON CONSUMER HEALTH

Milk is a food item which is used every day by almost all age groups and at every stage of life; that's why it can act as a carrier of essential nutrients to achieve sound health. Recently, food technology has been widely used in preparing dairy drinks with lower amounts of saturated fats; a lower saturated fat level offers cardiovascular health benefits, including reductions in blood lipid profiles, particularly cholesterol, LDL-cholesterol and triglycerides (Lopez-Huertas 2010). The effect of monounsaturated and polyunsaturated fatty acid–enriched diet on low-density and high-density lipoprotein cholesterol levels in healthy men and women indicated that a mixed diet high in monounsaturated fatty acids was equally effective as a diet rich in n-6 PUFA in lowering low-density lipoprotein cholesterol. The level of high-density lipoprotein cholesterol was slightly lowered in men but not in women in both diets (Mensink and Katan 1989). The influence of n-3 PUFA-enriched tomato juice and plain tomato juice on lipid profile, antioxidant biomarker status and cardiovascular disease (CVD) risk in 18 healthy women (35–55 years old) was compared by Garcia-Alonso et al. (2012). It was shown that intake of n-3 PUFA-enriched tomato juice for a short period did not change the serum lipid profile but an overall greater positive improvement of CVD risk factors was noticed, which indicated a promising synergistic action between tomato antioxidants and n-3 PUFA intake. A study on the influences of giving foods containing n-3 PUFAs either naturally (fresh fish, canned fish, flaxseed meal, canola oil) or from fortification with fish oil (margarine spread, milk, sausages, luncheon meat, French onion dip) on intake and blood and tissue ratios of 16 healthy males was conducted (Metcalf et al. 2003). Enrichment of novel foods with fish oil is a good option for enhancing the level of n-3 PUFAs in blood and tissue for consumer health.

9.8 MAJOR CHALLENGES ASSOCIATED WITH PUFA ENRICHMENT OF FOODS AND OPTIONS

PUFAs are unsaturated in nature and double bonds are present in abundance which, renders them extremely vulnerable to fat oxidation and unsuitable for adding into foods having long shelf life due to the development of objectionable fishy and rancid off-flavors (Anbudhasan et al. 2014). Formulating EFA-enriched products which are stable and convenient and have an acceptable flavor/taste within an affordable price range is really a challenging task for food producers. Food trials using numerous technologies and studies regarding health benefits and bioavailability are still underway and necessary before EFA supplementation in foods can attain the market on a worldwide scale (Ganesh and Hettiarachchy 2016). Loss of sensory qualities and decreased amounts of n-3 PUFAs, lipid oxidation and consequent reduced storage life are some major problems associated with the storage of enriched products (Cherian 2013). Rumen bio-hydrogenation, fatty acid esterification, capacity of cellular uptake and differences in addition of fatty acids are the main problems associated with easy transfer of dietary PUFA to milk fat; therefore, manipulation of the cow genome might be a good option for enhancing fatty acids greater than diet alone, but the physiology of the cow may be a challenge for this approach (Lanier and Corl 2015).

On the other hand, algal biomass is also as efficient as fish oil in DHA enrichment, although some plant-derived alternatives are also available for the enrichment of chicken. The use of antioxidants as food additives and plant-based sources like echiumand linseed oils can be helpful for reducing fishy flavors in marine-based sources. Fish oil–enriched food items and microencapsulated powder of fish oil must be stored for short periods, and especially packaged in one-portion units, to remove all factors increasing the oxidation of PUFAs because, owing to oxidation n-3 fatty acids can transform into extremely toxic peroxides and by-products of oxidation, which are responsible for various detrimental health effects including cancer etc. (Grune et al. 2001; Grundt et al. 2003; Kolanowski 2005). Consumption of a diet rich in n-3 PUFAs can be helpful in the deposition of fatty acids in broiler chickens, and fish oil can be a good source for the enrichment of poultry meat with EPA and DHA, but some sensory issues are related to this. The use of new techniques for masking the undesirable tastes and flavors would be a better option (Al-Khalifa 2015).

9.9 FUTURE PROSPECTS

For people who are non-vegetarian, n-3 PUFA-enriched food products provide a better way of consuming the essential fatty acids without any unwanted change in their dietary habits. Additionally, it was proved that the bioavailability of n-3 PUFAs from enriched foods is equivalent to fish oil capsule supplements (Kolanowski 2005; Ansorena and Astiasaran 2013). The need for new n-3 PUFA-enriched foods is supposed to undergo enhancement and diversification in the future, which will pose a challenging task to food producers and technologists. To obtain success, it is necessary for food technologists to understand the most important factors affecting lipid oxidation and sensory acceptability in their food products. New and more advanced technologies to produce delivery systems will most likely emerge in the coming years. For example, electrospray was recently used to produce nanostructures with encapsulated omega-3 PUFA, which had a higher oxidative stability than the neat oil (Torres-Giner et al. 2010), but these structures have not yet been evaluated in complex food emulsions. Also, the safety of using these structures in food products should be extensively studied as well.

9.10 CONCLUDING REMARKS

Polyunsaturated fatty acids are vital for human beings to live healthily. There are various natural sources to get EFAs in the diet, but during processing or preservation these get reduced. To compensate for processing losses, for maintaining a proper n-6/n-3 ratio and for improving the life quality of consumers, the enrichment of food products with n-3 PUFAs using the latest technologies and innovations must be carried out at a large scale.

REFERENCES

Alasalvar, C., Taylor, K. D. A., Zubcob, E., Shahidi, F. and Alexis, M. (2002). Differentiation of Cultured and Wild Sea Bass (*Dicentrarchuslabrax*): Total Lipid Content Fatty Acid and Trace Mineral Composition. *Food Chemistry*, 79(2):145–150.

Al-Khalifa, H. (2015). Production of Value-Added Poultry Meat: Enrichment with n-3 Polyunsaturated Fatty Acids. *World's Poultry Science Journal*, 71(2):319–326. DOI: 10.1017/S004393391500032X.

Anbudhasan, P., Surendraraj, A., Karkuzhali, S. and Ramasamy, D. (2014). Development of Omega 3 Fatty Acid Enriched Stable Functional Foods: Challenges and Options. *International Journal of Innovative Research and Review*, 2(1):1–13.

Ansorena, D. and Astiasarán, I. (2013). *Enrichment of Meat Products with Omega-3 Fatty Acids by Methods Other than Modification of Animal Diet*. Woodhead Publishing Limited, 299–318. DOI: 10.1533/9780857098863.3.299.

Arrigo, M. D., Hoz, L., Cambero, I., Lopez-Bote, C. J., Pin, C. and Ordonez, J. A. (2004). Production of n-3 Fatty Acid Enriched Pork Liver Pate. *Lebensm-Wiss u.-Technol*, 37:585–591.

Babuskin, S., Krishnan, K. R., Babu, P. A. S., Sivarajan, M. and Sukumar, M. (2014). Functional Foods Enriched with Marine Microalga *Nannochloropsisoculata* as a Source of w-3 Fatty Acids. *Food Technology and Biotechnology*, 52(3):292–299.

Barrow, C. J. and Wang, B. (2013). *Spray Drying and Encapsulation of Omega-3 Oils*. Woodhead Publishing Limited, 194–225. DOI: 10.1533/9780857098863.2.194.

Belarbi, E. H., Molina, E. and Chisti, Y. (2000). A Process for High and Scaleable Recovery of High Purity Eicosapentaenoic Acid Esters from Microalgae and Fish Oil. *Enzyme and Microbial Technology*, 26(7):516–529.

Beligon, V., Christophe, G., Fontanille, P. and Larroche, C. (2016). Microbial Lipids as Potential Sources to Food Supplements. *Current Opinion in Food Science*, 7:35–42.

Blanchard, H., Pedrono, F., Boulier-Monthean, N., Catheline, D., Rioux, V. and Legrand, P. (2013). Comparative Effects of Well-Balanced Diets Enriched in α-Linolenic or Linoleic Acids on LC-PUFA Metabolism in Rat Tissues. *Prostaglandins, Leukotrienes and Essential Fatty Acids*, 88(5):383–389.

Bourre, J. M. (2005). Where to Find Omega-3 Fatty Acids and How Feeding Animals with Diet Enriched in Omega-3 Fatty Acids to Increase Nutritional Value of Derived Products for Human: What Is Actually Useful?. *The Journal of Nutrition, Health and Aging*, 9(4):232–242.

Calder, P. C. (2013). *Nutritional Benefits of Omega-3 Fatty Acids*. Woodhead Publishing Limited, 3–26. DOI: 10.1533/9780857098863.1.3.

Certik, M., Adamechova, Z. and Guothova, L. (2013a). Simultaneous Enrichment of Cereals with Polyunsaturated Fatty Acids and Pigments by Fungal Solid State Fermentations. *Journal of Biotechnology*, 168(2):130–134.

Certik, M., Adamechova, Z., Hanusova, V. and Breierova, E. (2008). Biotechnology as a Useful Tool for Nutritional Improvement of Cereal-Based Materials Enriched with Polyunsaturated Fatty Acids and Pigments. *Actaagronomicahungarica*, 56(4):377–384. DOI: 10.1556/AAgr.56.2008.4.1.

Certik, M., Klempova, T., Guothova, L., Mihálik, D. and Kraic, J. (2013 b). Biotechnology for the Functional Improvement of Cereal-Based Materials Enriched with PUFA and Pigments. *European Journal of Lipid Science and Technology*, 115(11):1247–1256.

Chee, C. P., Gallaher, J. J., Djordjevic, D., Faraji, H., Mcclements, D. J., Decker, E. A., Hollender, R., Peterson, D. G., Roberts, R. F. and Coupland, J. N. (2005). Chemical and Sensory Analysis of Strawberry Flavoured Yogurt Supplemented with an Algae Oil Emulsion. *The Journal of Dairy Research*, 72(3):311–316.

Cherian, G. (2013). *Egg Enrichment with Omega-3 Fatty Acids*. Woodhead Publishing Limited, 288–298. DOI: 10.1533/9780857098863.3.288.

Chikwanha, O. C., Vahmani, P., Muchenje, V., Dugan, M. E. R. and Mapiye, C. (2017). Nutritional Enhancement of Sheep Meat Fatty Acid Profile for Human Health and Wellbeing. *Food Research International*. DOI: 10.1016/j.foodres.2017.05.005.

Cournarie, F., Savelli, M. P., Rosilio, V., Bretez, F., Vauthier, C., Grossiord, J. L. and Seiller, M. (2004). Insulin-Loaded w/o/w Multiple Emulsions: Comparison of the Performances of Systems Prepared with Medium-Chain-Triglycerides and Fish Oil. *European Journal of Pharmaceutics and Biopharmaceutics*, 58(3):477–482.

Dewhurst, R. J. and Moloney, A. P. (2013). *Modification of Animal Diets for the Enrichment of Dairy and Meat Products with Omega-3 Fatty Acids*. Woodhead Publishing Limited, 257–287. DOI: 10.1533/9780857098863.3.257.

Domingo, J. L. and Bocio, A. (2007). Levels of PCDD/PCDFs and PCbs in Edible Marine Species and Human Intake: A Literature Review. *Environment International*, 33(3):397–405.

Drusch, S. (2007). Sugar Beet Pectin: A Novel Emulsifying Wall Component for Microencapsulation of Lipophilic Good Ingredients by Spray-Drying. *Food Hydrocolloids*, 21(7):1223–1228.

Farrell, D. J. (1998). Enrichment of Hen Eggs with n-3 Long-Chain Fatty Acids and Evaluation of Enriched Eggs in Humans. *American Journal of Clinical Nutrition*, 68(3):538–544.

Feizollahi, E., Hadian, Z. and Honarvar, Z. (2018). Food Fortification with Omega-3 Fatty Acids; Microencapsulation as an Addition Method. *Current Nutrition and Food Science*, 14(2):90–103. DOI: 10.2174/1573401313666170728151350.

Food and Drug Administration (2000). Letter Regarding Dietary Supplement Health Claim for Omega-3 Fatty Acids and Coronary Heart Disease. http://www.fda.gov/.

Fraeye, I., Bruneel, C., Lemahieu, C., Buyse, J., Muylaert, K. and Foubert, I. (2012). Dietary Enrichment of Eggs with Omega-3 Fatty Acids: A Review. *Food Research International*, 48(2):961–969.

Fuentes, A., Fernandez-Segovia, I., Serraand, J. A. and Barat, J. M. (2010). Comparison of Wild Cultured Sea Bass (*Dicentrarchuslabrax*) Quality. *Food Chemistry*, 119(4):1514–1518.

Ganesan, B., Brothersen, C. and McMahon, D. J. (2014). Fortification of Foods with Omega-3 Polyunsaturated Fatty Acids. *Critical Reviews in Food Science and Nutrition*, 54(1):98–114. DOI: 10.1080/10408398.2011.578221.

Ganesh, V. and Hettiarachchy, N. S. (2016). A Review: Supplementation of Foods with Essential Fatty Acids—Can It Turn a Breeze without Further Ado? *Critical Reviews in Food Science and Nutrition*, 56(9):1417–1427. DOI: 10.1080/10408398.2013.765383.

Garcia-Alonso, F. J., Jorge-Vidal, V., Ros, G. and Periago, M. J. (2012). Effect of Consumption of Tomato Juice Enriched with n-3 Polyunsaturated Fatty Acids on the Lipid Profile, Antioxidant Biomarker Status, and Cardiovascular Disease Risk in Healthy Women. *European Journal of Nutrition*, 51(4):415–424. DOI: 10.1007/s00394-011-0225-0.

Garg, M. L., Wood, L. G., Singh, H. and Moughan, P. J. (2006). Means of Delivering Recommended Levels of Long Chain n-3 Polyunsaturated Fatty Acids in Human Diets. *Journal of Food Science*, 71(5):66–71.

Givens, d. I., Cottrill, B. R., Davies, M., Lee, P. A., Mansbridge, R. J. and Moss, A. R. (2000). Sources of n-3 Polyunsaturated Fatty Acids Additional to Fish Oil for Livestock Diets: A Review. *Nutrition Abstracts and Review Series B*, 70:3–19.

Gouin, S. (2004). Microencapsulation. *Trends in Food Science and Technology*, 15(7–8):330–347.

Grundt, H., Nilsen, D. W., Mansoor, M. A. and Nordoy, A. (2003). Increased Lipid Peroxidation During Long-Term Intervention with High Doses of n-3 Fatty Acids (PUFAs) Following an Acute Myocardial Infarction. *European Journal of Clinical Nutrition*, 57(6):793–800.

Grune, T., Kramer, K., Hoppe, P. P. and Siems, W. (2001). Enrichment of Eggs with n-3 Polyunsaturated Fatty Acids: Effects of Vitamin E Supplementation. *Lipids*, 36(8):833–838.

Heinzelmann, K., Franke, K., Velasco, J. and Márquez-Ruíz, G. (2000). Microencapsulation of Fish Oil by Freeze-Drying Techniques and Influence of Process Parameters on Oxidative Stability during Storage. *European Food Research and Technology*, 211(4):234–239.

Hennalu, F. S. and Norziah, M. H. (2011). Contribution of Microencapsulated n-3 PUFA Powder Toward Sensory and Oxidative Stability of Bread. *Journal of Food Processing and Preservation*, 35(5):596–604.

Hinriksdottir, H. H., Jonsdottir, V. L., Sveinsdottir, K., Martinsdottir, E. and Ramel, A. (2014). Bioavailability of Long-Chain n-3 Fatty Acids from Enriched Meals and from Microencapsulated Powder. *European Journal of Clinical Nutrition*, 1–5. DOI: 10.1038/ejcn.2014.250.

Hossain, M. A. (2011). Fish as Source of n-3 Polyunsaturated Fatty Acids (PUFAs), Which One Is Better-Farmed or Wild? *Advance Journal of Food Science and Technology*, 3(6):455–466.

Howe, P. R. C., Downing, J. A., Grenyer, B. F. S., Grigonis-Deane, E. M. and Bryden, W. L. (2002). Tuna Fishmeal as a Source of DHA for n-3 PUFA Enrichment of Pork, Chicken, and Eggs. *Lipids*, 37(11).

Iafelice, G., Caboni, M. F., Cubadda, R., Criscio, T. D., Trivisonno, M. C. and Marconi, E. (2008). Development of Functional Spaghetti Enriched with Long Chain Omega-3 Fatty Acids. *Cereal Chemistry Journal*, 85(2):146–151.

Jacobsen, C. (2010a). Enrichment of Foods with Omega-3 Fatty Acids: A Multidisciplinary Challenge. *Annals of the New York Academy of Sciences*, 1190:141–150. DOI: 10.1111/j.1749-6632.2009.05263.x.

Jacobsen, C. (2010b). Challenges When Developing Omega-3 Enriched Foods. *Oléagineux, Corps Gras, Lipides*, 17:251–258.

Jacobsen, C., Horn, A. F. and Nielsen, N. S. (2013a). *Enrichment of Emulsified Foods with Omega-3 Fatty Acids*. Woodhead Publishing Limited, 336–352. DOI: 10.1533/9780857098863.3.336.

Jakobsen, K. (1999). Dietary Modifications of Animal Fats: Status and Future Perspectives. *Lipid - Fett*, 101(12):475–483.

Justi, K. C., Hayashi, C., Visentainer, J. V., de Souza, N. E. and Matsushita, M. (2003). The Influence of Feed Supply Time on the Fatty Acid Profile of Nile Tilapia (*OreochomisNiloticus*) Fed on A Diet Enriched With n-3 Fatty Acids. *Food Chemistry*, 80(4):489–493.

Klaypradit, W. and Huang, Y. W. (2008). Fish Oil Encapsulation with Chitosan Using Ultrasonic Atomizer. *Lebensmittel-Wissenschaft und -Technologie*, 41(6):1133–1139.

Klinkesorn, U., Sophanodora, P., Chinachoti, P., Decker, E. A. and Mcclements, D. J. (2006). Characterization of Spray-Dried Tuna Oil Emulsified in Two-Layered Interfacial Membranes Prepared Using Electrostatic Layer-By-Layer Deposition. *Food Research International*, 39(4):449–457.

Kolanowski, W. (2005). Bioavailability of Omega-3 PUFA from Foods Enriched with Fish Oil: A Mini Review. *Polish Journal of Food and Nutrition Sciences*, 14/55(4):335–340.

Kolanowski, W. and Laufenberg, G. (2006). Enrichment of Food Products with Polyunsaturated Fatty Acids by fish Oil Addition. *European Food Research and Technology*, 222(3–4):472–477. DOI: 10.1007/s00217-005-0089-8.

Kolanowski, W., Swiderski, F. and Berger, S. (1999). Possibilities of Fish Oil Application for Food Products Enrichment with n-3 PUFA. *International Journal of Food Sciences and Nutrition*, 50(1):39–49.

Kris-Etherton, P. M., Harris, W. S., Appel, L. J. and Nutrition Committee (2003). Fish Consumption, Fish Oil, Omega-3 Fatty Acids, and Cardiovascular Disease. *Arteriosclerosis, Thrombosis, and Vascular Biology*, 23(2):20–30.

Kris-Etherton, P. M., Taylor, D. S., Yu-Poth, S., Huth, P., Moriarty, K., Fishell, V., Hargrove, R. L., Zhao, G. and Etherton, T. D. (2000). Polyunsaturated Fatty Acids in the Food Chain in the United States. *American Journal of Clinical Nutrition*, 71:179–188.

Kuratko, C., Abril, J. R., Hoffman, J. P. and Salem, N. (2013). *Enrichment of Infant Formula with Omega-3 Fatty Acids.* Woodhead Publishing Limited, 353–386. DOI: 10.1533/9780857098863.3.353.

Lanier, J. S. and Corl, B. A. (2015). Challenges in Enriching Milk Fat with Polyunsaturated Fatty Acids. *Journal of Animal Science and Biotechnology*, 6(1):26.

Lemahieu, C., Bruneel, C., Termote-Verhalle, R., Muylaert, K., Buyse, J. and Foubert, I. (2014). Effect of Different Microalgal n-3 PUFA Supplementation Doses on Yolk Color and n-3 LC-PUFA Enrichment in the Egg. *Algal Research*, 6:119–123.

Lopez-Ferrer, S., Baucells, M. D., Barroeta, A. C., Galobart, J. and Grashorn, M. A. (2001). n-3 Enrichment of Chicken Meat. 2. Use of Precursors of Long-Chain Polyunsaturated Fatty Acids: Linseed Oil. *Poultry Science*, 80(6):753–761.

Lopez-Huertas, E. (2010). Health Effects of Oleic Acid and Long Chain Omega-3 Fatty Acids (EPA and DHA) Enriched Milk: A Review of Intervention Studies. *Pharmacological Research*, 61(3):200–207.

Mantzioris, E., Cleland, L. G., Gibson, R. A., Neumann, M. A., Demasi, M. and James, M. J. (2000). Biochemical Effects of a Diet Containing Foods Enriched with n-3 Fatty Acids. *American Journal of Clinical Nutrition*, 72(1):42–48.

Mbatia, B., Mattiasson, B., Mulaa, F. and Adlercreutz, P. (2011). Strategies for the Enzymatic Enrichment of PUFA from Fish Oil. *European Journal of Lipid Science and Technology*, 113(6):717–723.

Mensink, R. P. and Katan, M. B. (1989). Effect of a Diet Enriched with Monounsaturated or Polyunsaturated Fatty Acids on Levels of Low-Density and High Density Lipoprotein Cholesterol in Healthy Women and Men. *The New England Journal of Medicine*, 321(7):436–441.

Metcalf, R. G., James, M. J., Mantzioris, E. and Cleland, L. G. (2003). A Practical Approach to Increasing Intakes of n-3 Polyunsaturated Fatty Acids: Use of Novel Foods Enriched with n-3 Fats. *European Journal of Clinical Nutrition*, 57(12):1605–1612. DOI: 10.1038/sj.ejcn.1601731.

Moharana, T. R., Byreddy, A. R., Puri, M., Barrow, C. and Rao, N. M. (2016). Selective Enrichment of Omega-3 Fatty Acids in Oils by Phospholipase A1. *PLoS One*, 11(3):e0151370. DOI: 10.1371/journal.

Nettleton, J. (1995). *Omega-3 Fatty Acids and Health*. Chapman & Hall, New York.

Patch, C. S., Tapsell, L. C., Mori, T. A., Meyer, B. J., Murphy, K. J., Mansour, J., Noakes, M., Clifton, P. M., Puddey, I. B., Beilin, L. J., Annison, G. and Howe, P. R. C. (2005). The Use of Novel Foods Enriched with Long-Chain n-3 Fatty Acids to Increase Dietary Intake: A Comparison of Methodologies Assessing Nutrient Intake. *Journal of the American Dietetic Association*, 105(12):1918–1926.

Petrovic, M., Gacic, M., Karacic, V., Gottstein, Z., Mazija, H. and Medic, H. (2012). Enrichment of Eggs in n-3 Polyunsaturated Fatty Acids by Feeding Hens with Different Amount of Linseed Oil in Diet. *Food Chemistry*, 135(3):1563–1568.

Qi, B., Fraser, T., Mugford, S., Dobson, G., Butler, J., Napier, J. A., Napier, J. A., Stobart, A. K. and Lazarus, C. M. (2004). Production of Very Long Chain Polyunsaturated Omega-3 and Omega-6 Fatty Acids in Plants. *Nature Biotechnology*, 22(6):739–745.

Rasti, B., Erfanian, A. and Selamat, J. (2017). Novel Nanoliposomal Encapsulated Omega-3 Fatty Acids and Their Applications in Food. *Food Chemistry*. DOI: 10.1016/j.foodchem.2017.03.089.

Sahena, F., Zaidul, I. S. M., Jinap, S., Saari, N., Jahurul, H. A., Abbas, K. A. and Norulaini, N. A. (2009). PUFAs in Fish: Extraction, Fractionation, Importance in Health. *Comprehensive Reviews in Food Science and Food Safety*, 8(2):59–74.

Sanders, T. A. B. (2000). Polyunsaturated Fatty Acids in the Food Chain in Europe. *American Journal of Clinical Nutrition*, 71(1):176–180.

Shahidi, F. and Wanasundra, U. N. (1998). Omega-3 Fatty Acid Concentrates Trends: Nutritional Aspect and Production Technologies. *Food SciTechnol*, 9:230–240.

Simopoulos, A., Leaf, A. and Salem, N. (2000). Workshop Statement on the Essentiality of and Recommended Dietary Intakes for Omega-6 and Omega-3 Fatty Acids. *Prostaglandins Leukot Essent Fatty Acids*, 63(3):119–121.

Surai, P. F. and Sparks, N. H. C. (2001). Designer Eggs: From Improvement of Egg Composition to Functional Food. *Trends in Food Science and Technology*, 12(1):7–16.

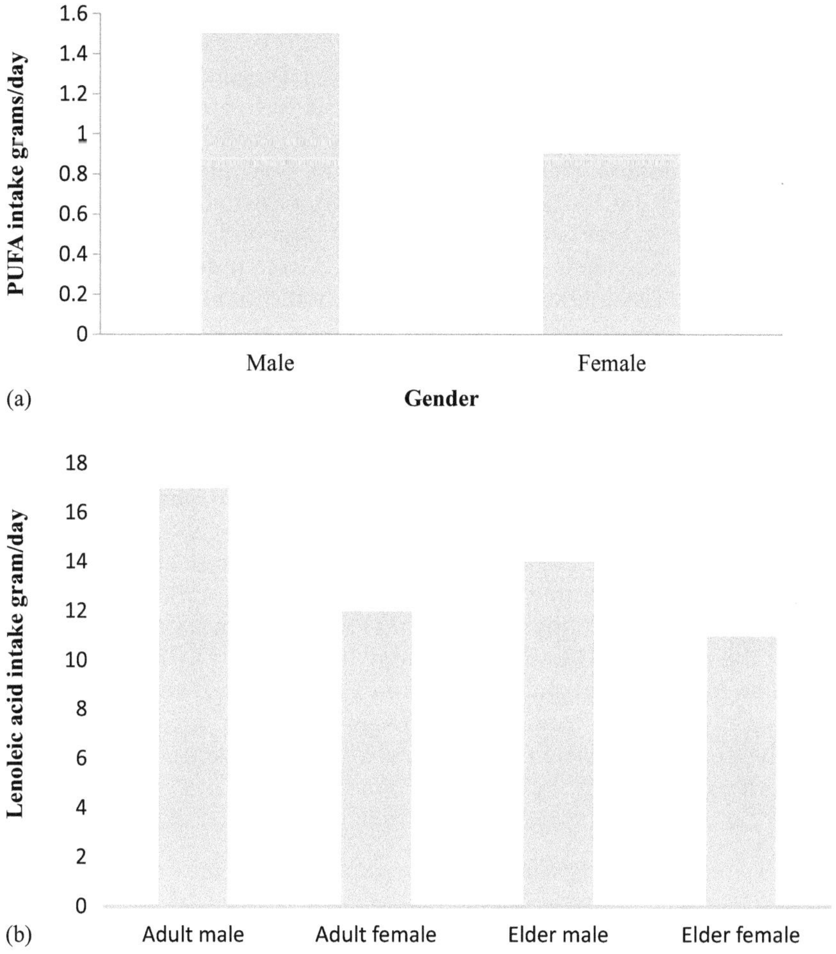

Figure 10.2 (a) Amount of PUFA intake for males and females; (b) amount of linoleic acid intake among the different age groups

CHD events more so than had been thought in the past (Nishida and Uauy 2009). In addition to CHD events and risk factors, there is an increased risk of metabolic syndrome components and diabetes. Therefore, this became a challenge for the food industry to remove TFA is a by-product of industrial processing (partial hydrogenation), usually in the form of partially hydrogenated vegetable oil (PHVO). The estimated intake of average daily TFA in most societies is low among adults. The experts recognize that there might be revision regarding the intake of recommended TFA (less than 1% E) which has not fully been taken into account. Thus, there is a need to protect substantial groups from having dangerously high intakes. This could lead to the need to remove partially hydrogenated fats and oils from the human food supply. In developed countries like the United States, the estimated average daily ruminant TFA intake is about 1.5 g for males and 0.9 g for females (Figure 10.3). Hence, the average intake for both males and females is 1.2 g, which corresponds to 0.5% E. If similar average intake values could be anticipated from industrially hydrogenated fat, then TFA intake from all sources should be no more than 1% E.

10.6 POLYUNSATURATED FATTY ACID INTAKE RECOMMENDATION TO PREVENT DEFICIENCY SYMPTOMS

The estimated convincing level to prevent deficiency symptoms is 2.5% E LA plus 0.5% E ALA. For lowering LDL and total cholesterol concentration, increasing HDL cholesterol concentration and decreasing risk of CHD events, the recommended level of PUFA consumption is minimal, i.e., 6% E based upon epidemiological studies. When tocopherol intake is low, the risk of lipid peroxidation may high (>11% E). This report was based on experimental studies (Elmadfa et al. 1989). However, this value was slightly different from a former recommendation given by WHO in 2003; hence the acceptable range for total PUFA (n-6 and n-3 fatty acid) lies between 6% and 11% E.

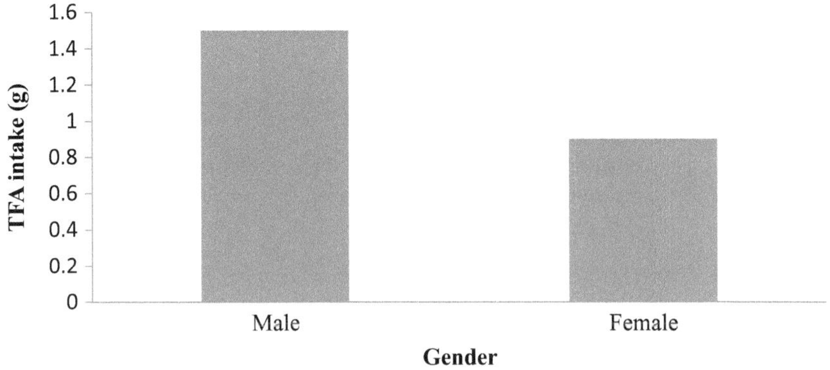

Figure 10.3 Trans-fatty acid intake in males and females in developed countries

10.7 CHALLENGES IN MEETING DIETARY RECOMMENDATIONS OF EFAS

The main consideration while meeting current recommendations is that today's dietary intake of PUFAs mainly consists of omega-6 FAs. Due to competition for the pathway enzymes (desaturases and elongases) in the metabolism of omega-3 and omega-6 FAs, the concentrations of these FAs are inversely affected (Cunnane 2003). Modern diets are high in refined carbohydrates, SFAs, trans-fatty acids and vegetable oil rich in omega-6 PUFAs but poor in omega-3 PUFAs (Simopoulos 2003). Vegetarian and vegan diets are even more concerning, having high LA combined with low EPA and DHA contents (Davis and Kris-Etherton 2003) and failing to meet the profile results, with low DHA concentration in plasma. Therefore, the inverse relation between omega-3 and omega-6 metabolism and PUFA profile of vegetarian and vegan diets is illustrated while attaining omega-3 FA recommendations (Rosell et al. 2005). In meeting recommendations with marine omega-3 FAs, low or no consumption of fish due to concern about environmental toxins, taste choices and ethical issues are other constraints. High amounts of environmental toxins such as mercury and polychlorinated biphenyls (PCBs), which can cause neurological damage to developing fetuses and young children, are reported in some fish (Guallar et al. 2002; Yamaguchi, Gazzard, Scholey and MacDonald 2003), providing a challenge in increasing the dietary intake of fish. Considering these issues, in March 2004, a joint advisory was issued for pregnant women, nursing mothers and young children by the Food and Drug Administration (FDA) and the Environmental Protection Agency (US Food and Drug Administration 2004). To minimize mercury exposure while benefitting from omega-3 PUFAs, three recommendations were made in the advisory: 1) do not eat fish containing high levels of mercury such king mackerel, shark fish, swordfish and tilefish; 2) eat 2 average meals (up to 340 g/week) of a variety of fish and shellfish (shrimp, canned light tuna, salmon, pollock and catfish); 3) for fish caught locally by family and friends, safety advisories should be checked. If no advice is available, one average meal (up to 170 g/week) of that fish should be consumed while avoiding further intake of any other fish during that week (US Food and Drug Administration 2004). Another concern for fish consumers is the level of PCBs, a toxin found to be higher in farm-cultivated than in wild salmon (Hites et al. 2004a). Including the skin in the analysis, studies confirmed a higher level of about 42 ppb of PCB concentrations compared to 3 ppb in wild salmon (Hites et al. 2004a, Carlson and Hites

2005). However, this level of PCBs did not present a significant risk per the FDA, as its limit in foods is 2000 ppb (FDA 2011). Additionally, these environmental toxins are further reduced by the preparation methods including removal of fish skin and cooking (Bayen et al. 2005). As the levels of these environmental toxins (mercury and PCBs) varied in different fishes, it is therefore suggested to consume a variety of fish, including wild as well as farmed fish, to minimize risk from toxins while maximizing health benefits. Non-consumption of fish by vegetarians and dislike of fish taste by some people is another challenge in meeting omega-3 PUFA recommendations. As only minimal amounts of EPA and small amounts of DHA are produced from ALA, other sources need to be used by vegans and people not eating fish to meet the dietary recommendations. Seaweed such as kelp, laver and wakame (EPA content ranging from 0.004 to 0.186 g/100 g portion) (Gebauer et al. 2006) can be a good option to increase the dietary intake of EPA by their incorporation into dishes such as sushi. DHA-rich microalgae and fish oil supplementation are some other options to increase the dietary intake of omega-3 PUFAs. In supplement/capsule form, DHA-rich microalgae provide 10–40% DHA (dwb) (Davis and Kris-Etherton 2003); therefore, only 100–300 mg/d supplementation is suggested. EPA and DHA amounts vary widely in fish oil capsules; up to 180 mg EPA and 120 mg DHA are commonly provided per capsule in the United States (Kris-Etherton, Harris and Appel 2002). Optimal ratios of EPA to DHA between 2:1 and 1:2 have been suggested recently for fish oil capsules, which can be attributed to 1) the apparent synergistic relation of EPA and DHA and 2) the lack of substantial evidence suggesting which one is the principal bioactive fatty acid (Harris 2004). Negligible mercury and PCB levels below the detection limit (Melanson et al. 2005) are other advantages supporting the use of fish oil supplementation.

10.8 CONCLUSION

Essential fatty acids are major components of the human diet, and we have to depend upon external dietary sources for the intake of these fatty acids. Several types of health issues are related to the deficiency of essential fatty acids in the human body. Therefore, an adequate amount of essential fatty acids is required to maintain the health of the human body. In this context, diet, age, gender and geographical location are the most important factors that influence the intake of the essential fatty acids. Apart from the intake of essential fatty acids, the food processing conditions always affect the bioavailability of the essential fatty acids. Based on the requirements and bioavailability, different amounts of fatty acids are required to maintain the health of a person. In conclusion, essential fatty acids are an important part of the human diet and sufficient intake of essential fatty acids is required according to age and gender.

REFERENCES

Amiri, M., Yousefnia, S., Forootan, F.S., Peymani, M., Ghaedi, K. and Esfahani, M.H.N. (2018). Diverse roles of fatty acid binding proteins (FABPs) in development and pathogenesis of cancers. *Gene*, 676, 171–183.

Authority, E.F.S. (2010). Scientific opinion on dietary reference values for fats, including saturated fatty acids, polyunsaturated fatty acids, monounsaturated fatty acids, trans fatty acids, and cholesterol. *EFSA Journal*, 8(3), 1461.

Bayen, S., Barlow, P., Lee, H.K. and Obbard, J.P. (2005). Effect of cooking on the loss of persistent organic pollutants from salmon. *Journal of Toxicology and Environmental Health, Part A*, 68(4), 253–265.

Berry, N.J. and Gribble, K.D. (2008). Breast is no longer best: Promoting normal infant feeding. *Maternal and Child Nutrition*, 4(1), 74–79.

Birch, Eileen E. et al. ((2010). The DIAMOND (DHA Intake And Measurement Of Neural Development) Study: A double-masked, randomized controlled clinical trial of the maturation of infant visual acuity as a function of the dietary level of docosahexaenoic acid. *The American Journal of Clinical Nutrition*, 91(4), 848–859.

Burdge, G.C. and Calder, P.C. (2005). Conversion of α-linolenic acid to longer-chain polyunsaturated fatty acids in human adults. *Reproduction, Nutrition, Development*, 45(5), 581–597.

Calder, P.C. (2015). Marine omega-3 fatty acids and inflammatory processes: Effects, mechanisms and clinical relevance. *Biochimica et Biophysica Acta (BBA)-Molecular and Cell Biology of Lipids*, 1851(4), 469–484.

Candela, C.G., López, L.B. and Kohen, V.L. (2011). Importance of a balanced omega 6/omega 3 ratio for the maintenance of health. Nutritional recommendations. *Nutricion Hospitalaria*, 26(2), 323–329.

Carlson, D.L. and Hites, R.A. (2005). Polychlorinated biphenyls in salmon and salmon feed: Global differences and bioaccumulation. *Environmental Science and Technology*, 39(19), 7389–7395.

Crawford, C.J. and Sigma Game Inc. (1999). *Slot Machine with Symbol Save Feature.* U.S. Patent 5,997,401.

Cunnane, S.C. (2003). Problems with essential fatty acids: Time for a new paradigm? *Progress in Lipid Research*, 42(6), 544–568.

Davis, B.C. and Kris-Etherton, P.M. (2003). Achieving optimal essential fatty acid status in vegetarians: Current knowledge and practical implications. *The American Journal of Clinical Nutrition*, 78(3), 640S–646S.

De Carvalho, C.C. and Caramujo, M.J. (2018). The various roles of fatty acids. *Molecules*, 23(10), 2583.

Eaton, S.B., Eaton III, S.B., Konner, M.J. and Shostak, M. (1996). An evolutionary perspective enhances understanding of human nutritional requirements. *The Journal of Nutrition*, 126(6), 1732–1740.

Elias, Sandra L. and Gibbons, Megan J. (2006). Nutritional foundation for pregnancy, childbirth and lactation. *Midwifery: Preparation for Practice*, 2, 415–430.

Elias, S.L. and Innis, S.M. (2001). Infant plasma trans, n–6, and n–3 fatty acids and conjugated linoleic acids are related to maternal plasma fatty acids, length of gestation, and birth weight and length. *The American Journal of Clinical Nutrition*, 73(4), 807–814.

Elmadfa, I., Schwalbe, P., Weidler, B. and Schlotzer, E. (1989). Effects of an intravenous application of dl-alpha-tocopheryl acetate on tocopherol status in man. *Annals of Nutrition and Metabolism*, 33(1), 1–6.

European Commission. Scientific Committee for Food. (1992). *Reports of the Scientific Committee for Food*. Commission of the European Communities Directorate General Information Market and Innovation.

Fabian, E. and Ibrahim, E. (2008). Nutritional situation of the elderly in the European Union: Data of the European Nutrition and Health Report. *Annals of Nutrition and Metabolism*, 52(Suppl. 1), 57–61.

Gebauer, S.K., Psota, T.L., Harris, W.S. and Kris-Etherton, P.M. (2006). n-3 fatty acid dietary recommendations and food sources to achieve essentiality and cardiovascular benefits. *The American Journal of Clinical Nutrition*, 83(6), 1526S–1535S.

Genzel-Boroviczényy, O., Wahle, J. and Koletzko, B. (1997). Fatty acid composition of human milk during the 1st month after term and preterm delivery. *European Journal of Pediatrics*, 156(2), 142–147.

Gibson, R.A. and Makrides, M. (1998). The role of long chain polyunsaturated fatty acids (LCPUFA) in neonatal nutrition. *Acta Paediatrica*, 87(10), 1017–1022.

Greenberg, J.A., Bell, S.J. and Van Ausdal, W. (2008). Omega-3 fatty acid supplementation during pregnancy. *Reviews in Obstetrics and Gynecology*, 1(4), 162.

Guallar, E., Sanz-Gallardo, M.I., Veer, P.V.T., Bode, P., Aro, A., Gómez-Aracena, J., Kark, J.D., Riemersma, R.A., Martín-Moreno, J.M. and Kok, F.J. (2002). Mercury, fish oils, and the risk of myocardial infarction. *New England Journal of Medicine*, 347(22), 1747–1754.

Harris, A. (2004). *Future Girl: Young Women in the Twenty-First Century*. Psychology Press, New York.

Hites, R.A., Foran, J.A., Carpenter, D.O., Hamilton, M.C., Knuth, B.A. and Schwager, S.J. (2004). Global assessment of organic contaminants in farmed salmon. *Science*, 303(5655), 226–229.

Innis, H.A. (2008). *The Bias of Communication*. University of Toronto Press.

Innis, S.M. (2005). Essential fatty acid transfer and fetal development. *Placenta*, 26, S70–S75.

Innis, S.M. and Friesen, R.W. (2008). Essential n– 3 fatty acids in pregnant women and early visual acuity maturation in term infants. *The American Journal of Clinical Nutrition*, 87(3), 548–557.

Institute of Medicine (US). Committee on Assuring the Health of the Public in the 21st Century. (2003). *The Future of the Public's Health in the 21st Century*. National Academy Press.

Jensen, C.L., Voigt, R.G., Prager, T.C., Zou, Y.L., Fraley, J.K., Rozelle, J.C., Turcich, M.R., Llorente, A.M., Anderson, R.E. and Heird, W.C. (2005). Effects of maternal docosahexaenoic acid intake on visual function and neurodevelopment in breastfed term infants. *The American Journal of Clinical Nutrition*, 82(1), 125–132.

Jéquier, E. and Tappy, L. (1999). Regulation of body weight in humans. *Physiological Reviews*, 79(2), 451–480.

Kang, J. (2005). Balance of Omega-6/Omega-3 Essential Fatty Acids Is Important for Health. *World Review of Nutrition and Dietetics*, 95, 93.

Kramer, M.S. and Kakuma, R. (2004). The optimal duration of exclusive breastfeeding. In: Pickering, L.K., Morrow, A.L. Ruiz-Palacios, G.M. and Schanl, R.J. (eds), *Protecting Infants Through Human Milk*, (pp. 63–77). Springer, Boston, MA.

Kris-Etherton, P.M., Harris, W.S. and, Appel, L.J. (2002). Fish consumption, fish oil, omega-3 fatty acids, and cardiovascular disease. *Circulation*, 106(21), 2747–2757.

Kuipers, R.S., Luxwolda, M.F., Offringa, P.J., Boersma, E.R., Dijck-Brouwer, D.J. and Muskiet, F.A. (2012). Fetal intrauterine whole body linoleic, arachidonic and docosahexaenoic acid contents and accretion rates. *Prostaglandins, Leukotrienes and Essential Fatty Acids*, 86(1–2), 13–20.

Lapillonne, A. and Jensen, C.L. (2009). Reevaluation of the DHA requirement for the premature infant. *Prostaglandins, Leukotrienes and Essential Fatty Acids*, 81(2–3), 143–150.

Lichtenstein, A.H. et al. (2006). Diet and lifestyle recommendations revision 2006: A scientific statement from the American Heart Association Nutrition Committee. *Circulation*, 114(1), 82–96.

Melanson, S.F., Lee Lewandrowski, E., Flood, J.G. and Lewandrowski, K.B. (2005). Measurement of organochlorines in commercial over-the-counter fish oil preparations: Implications for dietary and therapeutic recommendations for omega-3 fatty acids and a review of the literature. *Archives of Pathology and Laboratory Medicine*, 129(1), 74–77.

Mogensen, K.M. (2017). Essential fatty acid deficiency. *Practical Gastroenterology*, 41(6), 37–44.

Nishida, C. and Uauy, R. (2009). WHO Scientific Update on health consequences of trans fatty acids: Introduction. *European Journal of Clinical Nutrition*, 63, S1.

Punia, S., Sandhu, K.S., Siroha, A.K. and Dhull, S.B. (2019). Omega 3-metabolism, absorption, bioavailability and health benefits: A review. *PharmaNutrition*, 10, 100162.

Rosell, F., Bozser, O., Collen, P. and Parker, H. (2005). Ecological impact of beavers *Castor fiber* and *Castor canadensis* and their ability to modify ecosystems. *Mammal Review*, 35(3–4), 248–276.
Russo, G.L. (2009). Dietary n−6 and n−3 polyunsaturated fatty acids: From biochemistry to clinical implications in cardiovascular prevention. *Biochemical Pharmacology*, 77(6), 937–946.
Simopoulos, A.P. (1995). The Mediterranean food guide: Greek column rather than an Egyptian pyramid. *Nutrition Today*, 30(2), 54–61.
Simopoulos, A.P. (2003). Importance of the ratio of omega-6/omega-3 essential fatty acids: Evolutionary aspects. In: Simopoulos, A.P. and Cleland, L.G. (eds), *Omega-6/Omega-3 Essential Fatty Acid Ratio: The Scientific Evidence* (Vol. 92, pp. 1–22). Karger Publishers.
Simopoulos, A.P. (2006). Evolutionary aspects of diet, the omega-6/omega-3 ratio and genetic variation: Nutritional implications for chronic diseases. *Biomedicine and Pharmacotherapy*, 60(9), 502–507.
Simopoulos, A.P. (2009). Evolutionary aspects of the dietary omega-6:Omega-3 fatty acid ratio: Medical implications. In: Simopoulos, A.P. and De Meester, F. (eds), *A Balanced Omega-6/Omega-3 Fatty Acid Ratio, Cholesterol and Coronary Heart Disease* (Vol. 100, pp. 1–21). Karger Publishers.
Spector, A.A. (1999). Essentiality of fatty acids. *Lipids*, 34(1), S1–S3.
Szajewska, Hania, Horvath, Andrea and Koletzko, Berthold (2006). Effect of n-3 long-chain polyunsaturated fatty acid supplementation of women with low-risk pregnancies on pregnancy outcomes and growth measures at birth: A meta-analysis of randomized controlled trials. *The American Journal of Clinical Nutrition*, 83(6), 1337–1344.
Trumbo, P., Schlicker, S., Yates, A.A. and, Poos, M. (2002). Food and Nutrition Board of the Institute of Medicine, the National Academies. Dietary reference intakes for energy, carbohydrate, fiber, fat, fatty acids, cholesterol, protein and amino acids. *Journal of the American Dietetic Association*, 102(11), 1621–1630.
Uauy, R. and Dangour, A.D. (2009). Fat and fatty acid requirements and recommendations for infants of 0–2 years and children of 2–18 years. *Annals of Nutrition and Metabolism*, 55(1–3), 76.
Uauy, R., Mena, P. and Rojas, C. (2000). Essential fatty acids in early life: Structural and functional role. *Proceedings of the Nutrition Society*, 59(1), 3–15.
US Food and Drug Administration. (2004). What you need to know about mercury in fish and shellfish: 2004 EPA and FDA advice for women who might become pregnant, women who are pregnant, nursing mothers, young children. *US Food and Drug Administration (March 2004), Online: US Food and Drug Administration.*
Virsangbhai, C.K., Goyal, A., Tanwar, B. and Sihag, M.K. (2019). Potential health benefits of conjugated linoleic acid: An important functional dairy ingredient. *European Journal of Nutrition & Food Safety*, 11, 200–213.

Vorster, T.E., Klar, A., Soga, K. and Mair, R.J. (2005). Estimating the effects of tunneling on existing pipelines. *Journal of Geotechnical and Geoenvironmental Engineering*, 131(11), 1399–1410.
Yamaguchi, N., Gazzard, D., Scholey, G. and MacDonald, D. (2003). Concentrations and hazard assessment of PCbs, organochlorine pesticides and mercury in fish species from the upper Thames: River pollution and its potential effects on top predators. *Chemosphere*, 50(3), 265–273.
Yuhas, R., Pramuk, K. and Lien, E.L. (2006). Human milk fatty acid composition from nine countries varies most in DHA. *Lipids*, 41(9), 851–858.

11

Essential Fatty Acids and Their Health Benefits

Loveleen Sharma and Charanjiv Singh Saini

Contents

11.1 INTRODUCTION

In recent years there has been increased consumer interest in functional foods and active food components. Any food or dietary component that may be responsible for significant health benefits in addition to basic nutrition can be defined as a functional food (Kaur et al. 2014). The term "essential fatty acids" (EFAs) refers to those polyunsaturated fatty acids (PUFA) that must be provided by foods because they cannot be synthesized in the body yet are necessary for health. EFAs cannot be synthesized in the body and are essential for the survival of humans and other mammals; therefore, they have to be added to our diet and, thus, are essential. The ω-3 series derived from α-linolenic acid (ALA, 18:3) and the ω-6 series derived from cis-linoleic acid (LA, 18:2) are the two categories of naturally occurring EFAs in the body. Another sequence of fatty acids is derived from oleic acid (OA, 18:1 ω-9), but it is not an EFA since it can be derived from simple precursors in mammals; these precursors are known as PUFAs (polyunsaturated fatty acids) (Kliegman et al. 2015). There are two categories of PUFAs, ω-6 (omega-6 fatty acids) and ω-3 (omega-3 fatty acids), where ω-6 fatty acid is LA and ω-3 fatty acid is ALA (Punia et al. 2019). This review is centered on these two series of fatty acids and their metabolites. Metabolites of these two fatty acids (LA and ALA) are mainly required for regulating membrane function. The major importance of these long-chain fatty acids is in the brain, retina, liver, kidney, adrenal glands and gonads (Figure 11.1).

11.2 HEALTH BENEFITS

There is great promise that knowledge of fatty acid metabolism can be applied directly to human disease. In this report, we focus on those conditions with the strongest evidence but also point out conditions in which fatty acids are influential.

11.2.1 Inflammation

PUFAs and their products play important roles in the pathobiology of inflammation. The fatty acid content of cell membrane is responsible for the amount and type of PUFA released in response to inflammatory stimuli. Increased intake of the EFAs ALA and LA reduces the inflammatory response (Trebble et al. 2003). Kurlak et al. (1999) reported that

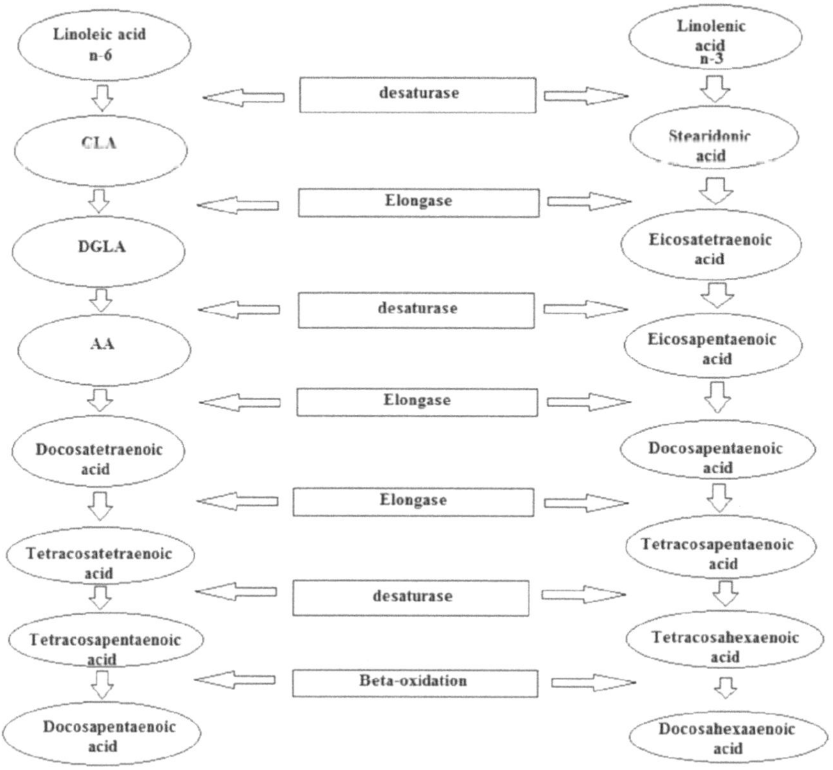

Figure 11.1 Metabolic process of n-6 and n-3 PUFA

supplementation of GLA, EPA and DHA exhibits anti-inflammatory actions that can be responsible for the reduction of formation of pro-inflammatory eicosanoids and cytokines and resulted in the enhanced production of beneficial eicosanoids. EPA and DHA also protect against oxidative stress and worked as terminators of free radical damage (Braverman and Moser 2012). EPA and DHA also inhibit the chronic overproduction of pro-inflammatory mediators, which might have adverse effects in certain conditions. EPA and DHA inhibit the effects of pro-inflammatory mediators, which may cause adverse effects in certain conditions. The immuno-modulatory activities of EPA and DHA are complex and include modification of lipid mediator responses, such as reduction in arachidonate-derived cytokines and increasing synthesis of resolvins via cyclo-oxygenases and lipoxygenases; phagocytosis enhancement; inhibition of

T-cell proliferation; and expression of major histocompatibility complexes I and II. It is reported that in exercise-trained men, C-reactive protein and tumor necrosis factor were reduced by intake of 2224 mg EPA and 2208 mg DHA. In the context of psychological stress, structural equation modeling of the influences of erythrocyte DHA, arachidonic acid and pro-inflammatory cytokines (interleukin-1, interleukin-6 and tumor necrosis factor) on perceived-stress-scale coping demonstrate that the benefits of DHA are largely direct effects unrelated to the effects of these cytokines. Despite these observations and significant basic research supporting its existence, a consistent relationship between EPA and DHA, inflammatory markers and oxidative stress in humans has been elusive.

11.2.2 Arthritis

Higher doses of EPA and DHA suppressed the arachidonate-derived eicosanoids, such as prostaglandin E_2 (Miles and Calder 2012). Randomized controlled trials conducted by Miles and Calder exhibited that EPA and DHA showed significant clinical improvement in rheumatoid arthritis. Long-time ingestion of fish oil reduces the expression of leukocyte inflammatory genes and decreases the cytokine production of leukocytes, which would improve inflammatory conditions such as rheumatoid arthritis (Baker et al. 2012). A separate meta-analysis of 17 randomized controlled trials using fish, seal, or krill oil (and one using flaxseed oil) showed significant benefit for inflammatory joint pain associated with rheumatoid arthritis, as well as inflammatory bowel disease and dysmenorrhea (Goldberg and Katz 2007).

11.2.3 Cardiovascular Disease

There is strong evidence that EPA and DHA prevent cardiovascular diseases (Mozaffarian and Wu 2011). Both EPA and DHA exhibited improvements in blood pressure, thrombosis, platelet reactivity, vascular reactivity and inflammation among others (Miles and Calder 2012). Trials conducted by different authors significantly exhibited the cardioprotective effect of fish consumption and omega-3 PUFA supplementation (Burr et al. 1989; Tavazzi et al. 2008). A meta-analysis of 19 observational studies (n = 45,637) and fatty acid biomarkers (total plasma, phospholipids, cholesterol esters and adipose tissue) using a continuous (per 1-SD increase) multivariable-adjusted analysis reported that EPA, docosapentaenoic acid (DPA) and DHA were associated with a 9–10% lower risk of fatal CHD

[RRs of 0.91 (95% CI, 0.82–1.00) for EPA, 0.90 (95% CI, 0.85–0.96) for DPA and 0.90 (95% CI, 0.84–0.96) for DHA]. In addition, the sum of EPA, DPA and DHA was associated with an 11% lower risk of fatal CHD (RR 0.89; 95% CI, 0.84–0.95). A 2016 analysis of two prospective cohort studies, the Nurses' Health Study (n = 83,349) and the Health Professionals Follow Up Study (n = 42,884), found that intake of marine-derived omega-3 PUFAs was associated with lower total mortality when comparing extreme quintiles of intake (HR 0.96; 95% CI, 0.93–1.00; p=0.002 for trend) (Wang et al. 2016). The GISSI-Heart Failure study showed reduced mortality and hospital admission for cardiovascular reasons in people with chronic heart failure already receiving recommended therapy when given 850 to 882 mg combined EPA and DHA (as ethyl esters in 1.2:1 ratio) per day (Marchioli et al. 2010). Risk reduction appeared to be due to antiarrhythmic effects. Low C20:4 n-3 (eicosatetraenoic acid, an EPA precursor) and high C18:1 n-7 (cis-vaccenic acid) have been associated with significantly higher mortality in chronic heart failure (Øie et al. 2011). In summary, current evidence strongly supports the roles of fatty acids, particularly EPA, DHA and ALA, in reducing the risk of cardiovascular disease. There is noteworthy evidence supporting their roles in heart failure, hypertension and atrial fibrillation as well.

11.2.4 Diabetes

High levels of linolenic acid, EPA and DHA are associated with a lower risk of type 2 diabetes in older people, revealed by analysis of plasma phospholipid (Djousse et al. 2012). On the contrary, in diabetics, blood glucose is positively associated with EPA and DHA intakes (Brostow et al. 2011). It was observed that dietary linoleic acid was positively but plasma phospholipid linoleic acid was inversely associated with type 2 diabetes risk in humans (Akinkuolie et al. 2011). Human studies exhibit that consumption of neither linolenic acid, EPA, nor DHA improves insulin sensitivity as compared to rodents (Wu et al. 2011). A meta-analysis showed that there is no association between the risk of developing type 2 diabetes and EPA, DHA, fish, or seafood intake (Wu et al. 2011).

11.2.5 Mental Health

Fatty acids are important for brain functions, as neuronal tissues have high lipid content. Sixty percent of lipid is present in the human brain, with DHA about 10% to 14%. DHA and arachidonic acid together constitute about 50% of the total fatty acids in neuronal membrane phospholipids. DHA also plays important roles in neuron size, neurogenesis, synapse formation and function, gene expression in brain, cognitive development, glucose transport and learning ability.

11.2.5.1 Schizophrenia

Studies exhibited a positive relationship between fatty acids and schizophrenia and semiology with sterol, anandamide, archiodonic acid and DHA. Significant augmentation of ALA-to-DHA biosynthesis was observed, due to which chronic risperidone treatment increased rat erythrocyte and prefrontal cortex DHA composition in a novel mechanism of action (McNamara et al. 2009). Erythrocyte analysis in 36 drug-free people with schizophrenia (vs. 36 controls) showed low arachidonic acid and DHA but an elevated n-6/n-3 ratio. After 3 months of typical antipsychotic drug treatment, these abnormalities normalized, consistent with crucial roles of fatty acids in both the etiology and treatment of schizophrenia (Sethom et al. 2010).

11.2.5.2 Depression

The evidence for essential fatty acid abnormalities in depression is strong. Despite design variations, 4 independent meta-analyses of controlled trials in patients with major depression and bipolar disorder found significant benefit from n-3 fatty acid treatment (Ross et al. 2007). EPA and DHA were found to be effective in mood disorders by some reviewers. n-3 fatty acid deficits and/or elevated serum, erythrocyte, adipose and brain phospholipid arachidonic acid/EPA or arachidonic acid/DHA ratios have been shown in a number of studies of unipolar and bipolar disorders. Mechanisms which can be possible for action for the use of ω-3 PUFAs as antidepressants include (a) release of inflammatory cytokines, which can provoke signs and symptoms as in major depressive disorder (Mischoulon and Fava 2000); (b) increase in serotonin 5-HT (hydroxytryptamine) transport by endothelial cells due to increase in membrane fluidity; (c) increased dopamine concentration and D (dopamine) 2 receptor binding due to increase in concentration of DHA in the frontal cortex (Hibbeln et al. 2007); and (d) mood alteration due to interaction with neuronal cell membrane receptors and second messengers (Mischoulon and Fava 2000). A recent meta-analysis from 1980 to 2014 using 35 random controlled trials exhibited that further random controlled trials are required to study populations with diagnosed or clinically significant depression of adequate duration using EPA-predominant ω-3 PUFA formulations (Hallahan et al. 2016).

11.2.5.3 Alzheimer's Disease and Dementia

Several epidemiological studies have exhibited that an increased risk of cognitive decline or dementia, especially for Alzheimer's disease, is linked with lower intakes of ω-3 PUFAs (Cole et al. 2009). Sufficient clinical evidence exists for ω-3s and prevention of Alzheimer's disease (MacLean et al. 2004). The primary component of membrane PLs in the brain is DHA, especially in the cerebral cortex, mitochondria, synaptosomes and synaptic vesicles (Connor 2000). Several studies exhibited the effects of ω-3s on dementia (Cole et al. 2009, Cunnane et al. 2009). The action mechanism of PUFAs on brain function include modifications to (a) the activity of membrane-bound enzymes, (b) membrane fluidity, (c) the production and activity of neurotransmitters, (d) the function of ion channels, (e) the number and affinity of receptors and (f) signal transduction, which controls the activity of neurotransmitters and neuronal growth factors (Yehuda et al. 2005). Decreases in lymphocyte proliferation are induced by ω-3 PUFAs, which could induce a TNFα production, natural killer cell activity and interleukin IL-1 and interleukin IL-2 production in humans (Boudrault et al. 2009, Singer and Richter-Heinrich 1991). Umbilical artery, umbilical vein FA compositions and neonatal neurological conditions in 317 term infants were investigated by Dijck-Brouwer et al. (2005), who reported that lower fetal DHA, AA and essential FA levels negatively influence the neurological condition of early postnatal infants. Lower intakes of essential fatty acids also lead to Huntington's disease, schizophrenia, high blood pressure and increased appetite signaling during adult life (Mathai et al. 2004). Several researchers reported that deficiency in essential FAs, especially ω-3 PUFAs, contributes to attention deficit hyperactivity disorder (ADHD) (Farooqui and Horrocks 2001; Ross et al. 2003). Consumption of ω-3 PUFAs significantly enhanced quality of life, sleep quality, ability to concentrate and hemoglobin levels in iron-deficient children (Yehuda et al. 2011). A positive effect was exerted by the consumption of fish and ω-3 PUFAs on cognitive health in older healthy adults (Cederholm 2017). Based on studies published during 2015–2016, it was reported that ω-3 supplementation could also benefit older adults with memory complaints/mild cognitive impairment and Alzheimer's disease (Cederholm 2017).

11.2.6 Maternal and Child Health

Maternal health during pregnancy and child health is associated with consumption of ω-3 PUFAs, as reported in plenty of epidemiological studies. Newberry et al. (2016) reported that ω-3 PUFAs affect length of gestation, preterm birth, birth weight, peripartum depression, gestational hypertension/preeclampsia, postnatal growth patterns, visual acuity, neurological development, cognitive development, autism spectrum disorder, ADHD, learning disorders, atopic dermatitis, allergies and respiratory disorders. Best et al. (2016) reported that meta-analysis based on clinical studies suggests the benefits of increased ω-3 PUFAs in the maternal diet and for outcomes of childhood allergic disease. A summary of research outcomes is shown in Table 11.1.

11.2.7 Visual and Neurological Development

Dietary ω-3 PUFAs play an important role in the development of the brain as reported by several studies (Clandinin et al. 1981; McNamara et al. 2018). High consumption of ω-3 PUFAs reduces delays in neurogenesis, neuronal migration, synaptic pruning, embryonic cortical plate expansion, brain glucose uptake and metabolism and can be responsible for impaired glutamate and monoamine synaptic function (McNamara et al. 2018). Within the visual system, DHA is a major structural component for retinal photoreceptors and cortical gray matter; hence, maturation of the visual system is assisted with DHA supplementation during pregnancy (Judge et al. 2007). Accumulation of DHA is rapid in neural cortex tissues and retinal membrane synapses during the second half of a pregnancy (Clandinin et al. 1981; Fleith and Clandinin 2005; Gould et al. 2013). Supplementation with ω-3 PUFAs during pregnancy has been observed to be positively associated with high-grade vocabulary comprehension, stereo-acuity, verbal intelligence quotient, receptive vocabulary and higher cognitive scores from infancy to nine years of age (Hibbeln et al. 2007).

11.2.8 Cancer

Over the past decade several experimental and epidemiological studies have exhibited that the ω-3 PUFAs reduce the risk of cancer. ω-3 PUFAs exert an anti-carcinogenic effect, whereas saturated and ω-6 PUFAs may enhance the chances of development (Prener et al. 1996). However, some systemic reviews report that there is insufficient evidence to suggest a significant association between ω-3 PUFAs and cancer incidence (MacLean et al. 2006). According to some studies, ω-3 PUFAs have been observed to affect different types of cancer, such as prostate, colon, breast,

Table 11.1 Research Findings of Omega-3 (ω-3) PUFA Effects on Maternal and Child Health

Findings	Total Participants	References
Overall greater gestation duration and infant size due to supplementation with ω-3s (600 mg DHA/day) in the last half of gestation	300 healthy pregnant women	Carlson et al. 2013
Supplementation with DHA (600 mg/day) from the first trimester until delivery effectively improved red cell membrane DHA eccentricity in pregnant women with type 2 diabetes and neonates and prevented reduction of maternal DHA during pregnancy	88 women with type 2 diabetes and 85 healthy women	Min et al. 2014
Supplementation with ω-3s (1,200 mg/day) did not exhibit any affirmation for reduction in arachidonic acid intake during pregnancy; lactation relevantly affected fat mass in offspring during the first year of life	170 healthy pregnant women before the 15th week of gestation, 18–43 years of age	Hauner et al. 2012
Consumption of functional foods rich in DHA during pregnancy could be beneficial with respect to infant body composition at birth and insulin sensitivity	37 healthy pregnant women, mid-pregnancy (20–24 weeks)	Courville et al. 2011
Repeated preterm birth is reduced by the regular intake of fish (up to 3 meals/week) prior to 22 weeks of gestation	852 randomized women	Klebanoff et al. 2011
No effect was observed with supplementation with EPA-rich fish oil (1,060 mg EPA plus 274 mg DHA) and DHA-rich fish oil (900 mg DHA plus 180 mg EPA) on depressive symptoms during pregnancy or postpartum	118 pregnant women at risk for depression (EPDS score of 9–19 or a history of depression) in early pregnancy	Mozurkewich et al. 2013

lung, colorectal, ovarian, pancreatic, skin and stomach (Takezaki et al. 2003; Kato et al. 1997). Also, ω-3 PUFAs have been observed to enhance the efficacy and tolerability of chemotherapy (Mocellin et al. 2017). It was observed that ω-3 PUFAs can inhibit tumor cell growth by reducing cyclooxygenase levels and hence could act as a natural COX inhibitor (Kobayashi et al. 2006). Berquin et al. (2007) reported that a ratio of ω-6 to ω-3 of lower than 5 was effective in slowing cancer progression. Recently it was reported that combining ω-3 with 1α, 25-dihydroxy-vitamin D3 significantly enhanced cell apoptosis in breast cancer cell lines (Yang et al. 2017). Shahverdi and Niknam (2017) suggested that flaxseed exerts an anti-proliferative activity in breast cancer patients and recommended 25 g per day of flax for premenopausal women. Supplementation with ω-3 to cancer patients undergoing radiotherapy helped them to maintain body weight and helped them to tolerate the anticancer treatment (Pedrazzoli et al. 2017). Finocchiaro et al. (2012) observed a significant increase in body weight as well as effective antioxidative and anti-inflammatory effects of ω-3 PUFAs.

11.3 CONCLUSION

Trials with clinical end points and research should be focused on understanding the mechanisms of essential fatty acids by which health benefits in coronary diseases can be obtained. Essential fatty acids exhibited positive health benefits on cardiovascular disease (atrial fibrillation, atherosclerosis, thrombosis, inflammation and sudden cardiac death), diabetes, cancer, depression and various mental illnesses, age-related cognitive decline, periodontal disease and rheumatoid arthritis. A number of techniques are used to generate food products enriched with n-3 fatty acids from either animal or plant sources. Microencapsulated foods are the future foods for the supplementation of essential fatty acids. Due to these techniques, a wide variety of products have been launched on the market, such as "Mother's Horlicks," "Cadbury Bournvita Li'l Champs" (with DHA), "Junior's Horlicks," "Omega Milk," etc., which can be consumed to gain the health benefits of essential fatty acids. Regardless of these impressive achievements, there are some limitations to the availability, quality and acceptability of enriched food products. Therefore, future research should be focused on the process of production of transgenic animals and plants which would be rich in essential fatty acids or which can convert n-6 to n-3 fatty acids.

REFERENCES

Akinkuolie, A.O., Ngwa, J.S., Meigs, J.B. and Djoussé, L. 2011. Omega-3 polyunsaturated fatty acid and insulin sensitivity: A meta-analysis of randomized controlled trials. *Clinical Nutrition* 30(6):702–707.

Baker, K.R., Matthan, N.R., Lichtenstein, A., Niu, J., Guermazi, A., et al. 2012. Association of plasma n-6 and n-3 polyunsaturated fatty acids with synovitis in the knee: The MOST study. *Osteoarthritis and Cartilage* 20(5):382–387.

Berquin, I.M., Min, Y., Wu, R., Wu, J., Perry, D. et al. 2007. Modulation of prostate cancer genetic risk by omega-3 and omega-6 fatty acids. *The Journal of Clinical Investigation* 117(7):1866–1875.

Best, K.P., Gold, M., Kennedy, D., Martin, J. and Makrides, M. 2016. Omega-3 long-chain PUFA intake during pregnancy and allergic disease outcomes in the offspring: A systematic review and meta-analysis of observational studies and randomized controlled trials. *The American Journal of Clinical Nutrition* 103(1):128–143.

Boudrault, C., Bazinet, R.P. and Ma, D.W. 2009. Experimental models and mechanisms underlying the protective effects of n-3 polyunsaturated fatty acids in Alzheimer's disease. *The Journal of Nutritional Biochemistry* 20(1):1–10.

Braverman, N.E. and Moser, A.B. 2012. Functions of plasmalogen lipids in health and disease. *Biochimica et Biophysica Acta (BBA)-Molecular Basis of Disease* 1822(9):1442–1452.

Brostow, D.P., Odegaard, A.O., Koh, W.P., Duval, S., Gross, M.D., Yuan, J.M. and Pereira, M.A. 2011. Omega-3 fatty acids and incident type 2 diabetes: The Singapore Chinese Health Study. *The American Journal of Clinical Nutrition* 94(2):520–526.

Burr, M.L., Fehily, A.M., Gilbert, J.F., Rogers, S., Holliday, R.M., Sweetnam, P.M., Elwood, P.C. and Deadman, N.M. 1989. Effects of changes in fat, fish and fibre intakes on death and myocardial reinfarction: Diet and Reinfarction Trial. *Lancet* 2(8666):757–761.

Carlson, S.E., Colombo, J., Gajewski, B.J., Gustafson, K.M., Mundy, D., Yeast, J., Georgieff, M.K., Markley, L.A., Kerling, E.H. and Shaddy, D.J. 2013. DHA supplementation and pregnancy outcomes. *American Journal of Clinical Nutrition* 97(4):808–815.

Cederholm, T. 2017. Fish consumption and omega-3 fatty acid supplementation for prevention or treatment of cognitive decline, dementia or Alzheimer's disease in older adults: Any news? *Current Opinion in Clinical Nutrition and Metabolic Care* 20(2):104–109.

Clandinin, M.T., Chappell, J.E., Heim, T., Swyer, P.R. and Chance, G.W. 1981. Fatty acid utilization in perinatal de novo synthesis of tissues. *Early Human Development* 5(4):355–366.

Cole, G.M., Ma, Q.L. and Frautschy, S.A. 2009. Omega-3 fatty acids and dementia. *Prostaglandins, Leukotrienes and Essential Fatty Acids* 81(2–3):213–221.

Connor, W.E. 2000. Importance of n-3 fatty acids in health and disease. *The American Journal of Clinical Nutrition* 71(1):171S–175S.

Courville, A.B., Harel, O. and Lammi-Keefe, C.J. 2011. Consumption of a DHA-containing functional food during pregnancy is associated with lower infant ponderal index and cord plasma insulin concentration. *British Journal of Nutrition* 106(2):208–212.

Cunnane, S.C., Plourde, M., Pifferi, F., Bégin, M., Féart, C. and Barberger-Gateau, P. 2009. Fish, docosahexaenoic acid and Alzheimer's disease. *Progress in Lipid Research* 48(5):239–256.

Dijck-Brouwer, D.J., Hadders-Algra, M., Bouwstra, H., Decsi, T., Boehm, G., Martini, I.A., Boersma, E.R. and Muskiet, F.A. 2005. Lower fetal status of docosahexaenoic acid, arachidonic acid and essential fatty acids is associated with less favorable neonatal neurological condition. *Prostaglandins, Leukotrienes and Essential Fatty Acids* 72(1):21–28.

Djousse, L., Akinkuolie, A.O., Wu, J.H., Ding, E.L. and Gaziano, J.M. 2012. Fish consumption, omega-3 fatty acids and risk of heart failure: A meta-analysis. *Clinical Nutrition* 31(6):846–853.

Farooqui, A.A. and Horrocks, L.A. 2001. Plasmalogens, phospholipase A 2, and docosahexaenoic acid turnover in brain tissue. *Journal of Molecular Neuroscience* 16(2–3):263–272.

Finocchiaro, C., Segre, O., Fadda, M., Monge, T., Scigliano, M. et al. 2012. Effect of n-3 fatty acids on patients with advanced lung cancer: A double-blind, placebo-controlled study. *British Journal of Nutrition* 108(2):327–333.

Fleith, M. and Clandinin, M.T. 2005. Dietary PUFA for preterm and term infants: Review of clinical studies. *Critical Reviews in Food Science and Nutrition* 45(3):205–229.

Goldberg, R.J. and Katz, J. 2007. A meta-analysis of the analgesic effects of omega-3 polyunsaturated fatty acid supplementation for inflammatory joint pain. *Pain* 129(1–2):210–223.

Gould, J.F., Smithers, L.G. and Makrides, M. 2013. The effect of maternal omega-3 (n-3) LCPUFA supplementation during pregnancy on early childhood cognitive and visual development: A systematic review and meta-analysis of randomized controlled trials. *The American Journal of Clinical Nutrition* 97(3):531–544.

Hallahan, B., Ryan, T., Hibbeln, J.R., Murray, I.T., Glynn, S., Ramsden, C.E., SanGiovanni, J.P. and Davis, J.M. 2016. Efficacy of omega-3 highly unsaturated fatty acids in the treatment of depression. *The British Journal of Psychiatry* 209(3):192–201.

Hauner, H., Much, D., Vollhardt, C., Brunner, S., Schmid, D., et al. 2012. Effect of reducing the n-6: n-3 long-chain PUFA ratio during pregnancy and lactation on infant adipose tissue growth within the first year of life: An open-label randomized controlled trial. *American Journal of Clinical Nutrition* 95(2):383–394.

Hibbeln, J.R., Davis, J.M., Steer, C., Emmett, P., Rogers, I., Williams, C. and Golding, J. 2007. Maternal seafood consumption in pregnancy and neurodevelopmental outcomes in childhood (ALSPAC study): An observational cohort study. *Lancet* 369(9561):578–585.

Judge, M.P., Harel, O. and Lammi-Keefe, C.J. 2007. A docosahexaenoic acid: Functional food during pregnancy benefits infant visual acuity at four but not six months of age. *Lipids* 42(2):117–122.

Kato, I., Akhmedkhanov, A., Koenig, K., Toniolo, P.G., Shore, R.E. and Riboli, E. 1997. Prospective study of diet and female colorectal cancer: The New York University Women's Health Study. *Nutrition and Cancer* 28(3): 276–281.

Kaur, N., Chugh, V. and Gupta, A.K. 2014. Essential fatty acids as functional components of foods: A review. *Journal of Food Science and Technology* 51(10):2289–2303.

Klebanoff, M.A., Harper, M., Lai, Y., Thorp Jr, J., Sorokin, Y. et al. 2011. Fish consumption, erythrocyte fatty acids, and preterm birth. *Obstetrics and Gynecology* 117(5):1071.

Kliegman, R.M., Stanton, B., Geme, J.S., Schor, N.F. and Behrman, R.E., 2015. *Nelson Textbook of Pediatrics: Growth and Development*. Elsevier Health Sciences.

Kobayashi, N., Barnard, R.J., Henning, S.M., Elashoff, D., Reddy, S.T. et al. 2006. Effect of altering dietary ω-6/ω-3 fatty acid ratios on prostate cancer membrane composition, cyclooxygenase-2, and prostaglandin E2. *Clinical Cancer Research* 12(15):4662–4670.

Kurlak, L.O. and Stephenson, T.J. 1999. Plausible explanations for effects of long chain polyunsaturated fatty acids (LCPUFA) on neonates. *Archives of Disease in Childhood-Fetal and Neonatal Edition* 80(2):148–154.

MacLean, C.H., Mojica, W.A., Morton, S.C., Pencharz, J., Garland, R.H., Tu, W., Newberry, S.J., Jungvig, L.K., Grossman, J., Khanna, P. and Rhodes, S., 2004. Effects of omega-3 fatty acids on lipids and glycemic control in type II diabetes and the metabolic syndrome and on inflammatory bowel disease, rheumatoid arthritis, renal disease, systemic lupus erythematosus, and osteoporosis: Summary. In: *AHRQ Evidence Report Summaries*. Agency for Healthcare Research and Quality (US), Rockville, MD.

MacLean, C.H., Newberry, S.J., Mojica, W.A., Khanna, P., Issa, A.M., Suttorp, M.J., Lim, Y.W., Traina, S.B., Hilton, L., Garland, R. and Morton, S.C. 2006. Effects of omega-3 fatty acids on cancer risk: A systematic review. *JAMA* 295(4):403–415.

Marchioli, R., Silletta, M.G., Levantesi, G., Pioggiarella, R. and Tognoni, G. 2010. N-3 polyunsaturated fatty acids in heart failure: Mechanisms and recent clinical evidence. *Cellular and Molecular Biology* 56(1):110–130.

Mathai, M.L., Soueid, M., Chen, N., Jayasooriya, A.P., Sinclair, A.J., Wlodek, M.E., Weisinger, H.S. and Weisinger, R.S. 2004. Does perinatal ω-3 polyunsaturated fatty acid deficiency increase appetite signaling? *Obesity Research* 12(11):1886–1894.

McNamara, R.K., Able, J.A., Jandacek, R., Rider, T. and Tso, P. 2009. Chronic risperidone treatment preferentially increases rat erythrocyte and prefrontal cortex omega-3 fatty acid composition: Evidence for augmented biosynthesis. *Schizophrenia Research* 107(2–3):150–157.

McNamara, R.K., Asch, R.H., Lindquist, D.M. and Krikorian, R. 2018. Role of polyunsaturated fatty acids in human brain structure and function across the lifespan: An update on neuroimaging findings. *Prostaglandins, Leukotrienes and Essential Fatty Acids* 136:23–34.

Miles, E.A. and Calder, P.C. 2012. Influence of marine n-3 polyunsaturated fatty acids on immune function and a systematic review of their effects on clinical outcomes in rheumatoid arthritis. *British Journal of Nutrition* 107:171–184.

Min, Y., Djahanbakhch, O., Hutchinson, J., Bhullar, A.S., Raveendran, M., Hallot, A., Eram, S., Namugere, I., Nateghian, S. and Ghebremeskel, K. 2014. Effect of docosahexaenoic acid–enriched fish oil supplementation in pregnant women with type 2 diabetes on membrane fatty acids and fetal body composition: Double-blinded randomized placebo-controlled trial. *Diabetic Medicine* 31(11):1331–1340.

Mischoulon, D. and Fava, M. 2000. Docosahexanoic acid and ω-3 fatty acids in depression. *Psychiatric Clinics of North America* 23(4):785–794.

Mocellin, M.C., de Quadros Camargo, C., de Souza Fabre, M.E. and de Moraes Trindade, E.B.S. 2017. Fish oil effects on quality of life, body weight and free fat mass change in gastrointestinal cancer patients undergoing chemotherapy: A triple blind, randomized clinical trial. *Journal of Functional Foods* 31:113–122.

Mozaffarian, D. and Wu, J.H. 2011. Omega-3 fatty acids and cardiovascular disease: Effects on risk factors, molecular pathways, and clinical events. *Journal of the American College of Cardiology* 58(20):2047–2067.

Mozurkewich, E.L., Clinton, C.M., Chilimigras, J.L., Hamilton, S.E., Allbaugh, L.J., Berman, D.R., Marcus, S.M., Romero, V.C., Treadwell, M.C., Keeton, K.L. and Vahratian, A.M. 2013. The mothers, omega-3, and mental health study: A double-blind, randomized controlled trial. *American Journal of Obstetrics and Gynecology* 208:313–316.

Newberry, S.J., Chung, M., Booth, M., Maglione, M.A., Tang, A.M. et al. 2016. Omega-3 fatty acids and maternal and child health: An updated systematic review. *Evidence Report/Technology Assessment* 224:821–826.

Øie, E., Ueland, T., Dahl, C.P., Bohov, P., Berge, C., Yndestad, A., Gullestad, L., Aukrust, P. and Berge, R.K. 2011. Fatty acid composition in chronic heart failure: Low circulating levels of eicosatetraenoic acid and high levels of vaccenic acid are associated with disease severity and mortality. *Journal of Internal Medicine* 270(3):263–272.

Pedrazzoli, P., Cereda, E., Cappello, S., Colombo, S., Klersy, C., et al. 2017. Nutritional counseling with or without systematic use of oral nutritional supplements in head and neck cancer patients undergoing radiotherapy. *Journal of Clinical Oncolology* 35:98–102.

Prener, A., Storm, H.H. and Nielsen, N.H. 1996. Cancer of the male genital tract in Circumpolar Inuit. *Acta Oncologica* 35(5):589–593.

Punia, S., Sandhu, K.S., Siroha, A.K. and Dhull, S.B. 2019. Omega 3 metabolism, absorption, bioavailability and health benefits: A review. *Pharma Nutrition*, 100162.

Ross, B.M., McKenzie, I., Glen, I. and Bennett, C.P.W. 2003. Increased levels of ethane, a non-invasive marker of n-3 fatty acid oxidation, in breath of children with attention deficit hyperactivity disorder. *Nutritional Neuroscience* 6(5):277–281.

Ross, B.M., Seguin, J. and Sieswerda, L.E. 2007. Omega-3 fatty acids as treatments for mental illness: Which disorder and which fatty acid? *Lipids in Health and Disease* 6:21.

Sethom, M.M., Fares, S., Bouaziz, N., Melki, W., Jemaa, R., Feki, M., Hechmi, Z. and Kaabachi, N. 2010. Polyunsaturated fatty acids deficits are associated with psychotic state and negative symptoms in patients with schizophrenia. *Prostaglandins, Leukotrienes and Essential Fatty Acids* 83(3):131–136.

Shahverdi, E. and Niknam, R. 2017. Flaxseed: A good idea for reduction of mastalgia (breast pain) and risk of breast cancer. *Focus on Medical Sciences Journal* 3:1–2.

Singer, P. and Richter-Heinrich, E. 1991. Stress and fatty liver—Possible indications for dietary long-chain n-3 fatty acids. *Medical Hypotheses* 36(1):90–94.

Takezaki, T., Inoue, M., Kataoka, H., Ikeda, S., Yoshida, M., Ohashi, Y., Tajima, K. and Tominaga, S. 2003. Diet and lung cancer risk from a 14-year population-based prospective study in Japan: With special reference to fish consumption. *Nutrition and Cancer* 45(2):160–167.

Tavazzi, L., Maggioni, A.P., Marchioli, R., Barlera, S., Franzosi, M.G., Latini, R., Lucci, D., Nicolosi, G.L., Porcu, M., Tognoni, G. and Gissi-HF Investigators. 2008. Effect of n-3 polyunsaturated fatty acids in patients with chronic heart failure (the GISSI-HF trial): A randomised, double-blind, placebo-controlled trial. *Lancet* 372(9645):1223–1230.

Trebble, T., Arden, N.K., Stroud, M.A., Wootton, S.A., Burdge, G.C., Miles, E.A., Ballinger, A.B., Thompson, R.L. and Calder, P.C. 2003. Inhibition of tumour necrosis factor-α and interleukin 6 production by mononuclear cells following dietary fish-oil supplementation in healthy men and response to antioxidant co-supplementation. *British Journal of Nutrition* 90(2):405–412.

Wang, D.D., Li, Y., Chiuve, S.E., Stampfer, M.J., Manson, J.E., Rimm, E.B., Willett, W.C. and Hu, F.B. 2016. Association of specific dietary fats with total and cause-specific mortality. *JAMA Internal Medicine* 176(8):1134–1145.

Wu, S., Liang, J., Zhang, L., Zhu, X., Liu, X. and Miao, D. 2011. Fish consumption and the risk of gastric cancer: Systematic review and meta-analysis. *BMC Cancer* 11:26.

Yang, J., Zhu, S., Lin, G., Song, C. and He, Z. 2017. Vitamin D enhances omega-3 polyunsaturated fatty acids-induced apoptosis in breast cancer cells. *Cell Biology International* 41(8):890–897.

Yehuda, S., Rabinovitz, S. and Mostofsky, D.I. 2005. Essential fatty acids and the brain: From infancy to aging. *Neurobiology of Aging* 26:98–102.

Yehuda, S., Rabinovitz-Shenkar, S. and Carasso, R.L. 2011. Effects of essential fatty acids in iron deficient and sleep-disturbed attention deficit hyperactivity disorder (ADHD) children. *European Journal of Clinical Nutrition* 65(10):1167–1169.

12

Dietary Fatty Acids

A Key Requirement for Healthy Life with Special Reference to Their Future Prospects

Ajay Singh, Ramandeep Kaur, Pradyuman Kumar and Ashish Kumar Singh

Contents

12.1 INTRODUCTION

Lipids are biologically active substances which are important constituents of living cells and perform several functions in human body, such as providing energy, maintaining the structural integrity of cell membranes and participating in signaling pathways. On oxidation, 1 g of fat in the diet releases 9 kilocalories of energy. They are non-polar in nature and soluble in many organic solvents (such as ether, alcohol and chloroform) (Subramaniam et al. 2011). Based on a chemically functional backbone component, lipids contain a broad range of substances such as fatty acids, waxes, sterols, phospholipids and others (Christie and Han 2003). These (also known as triglycerides) are made up of a combination of one molecule of glycerol with three fatty acids. Among these units, fatty acids are a key source of fat in the diet. These are carbon chains with a methyl group (-CH_3) at one end and a carboxyl group (-COOH) at the other end of the molecule. Furthermore, fatty acids are categorized into saturated (no double bonds), monounsaturated (1 double bond), and polyunsaturated (2 or more double bonds) and trans fatty acids (Figure 12.1) (Das 2006). Polyunsaturated fatty acids (PUFAs) are also considered as essential fatty acids (EFAs), thus to be taken through the diet. EFAs are of two types; the omega-6 (ω-6) series (derived from linoleic acid) and the omega-3 (ω-3) series (derived from α-linolenic acid) (Punia et al. 2019). There is another series of fatty acids, defined as omega-9 (ω-9) fatty acids (derived from oleic acid). The human body metabolizes all types of fatty acids through same kind of enzymes (*desaturase* and *elongase*) to their respective pathways. Various studies reported the beneficial effects of ω-3, ω-6 and ω-9 fatty acids in the prevention of cardiovascular, neurological and neurodegenerative disorders (Dyall 2011; Denis et al. 2015; Gonçalves-de-Albuquerque et al. 2016). Studies have also shown that a low intake of dietary EFAs is associated with the occurrence of various problems such as poor fetal development, increased inflammatory and cardiovascular diseases and also increased risk of Alzheimer's disease (Gonçalves-de-Albuquerque et al. 2016). However, diets with more intake of saturated fatty acids (SFAs) and trans fatty acids (TFAs) have adverse effects on the human body and initiate various diseases (Morenga and Montez 2017). The TFAs (produced during the hydrogenation process) largely used by food industries and their presence in food have become a major health concern for consumers. Since many companies are focusing on the development of modified fats (tailor-made fats) and oils that contain high levels of unsaturated fatty acids (USFAs) to permit the exclusion of the requirement for hydrogenation and subsequently, the production of TFAs (Knauf and Del Vecchio 1998).

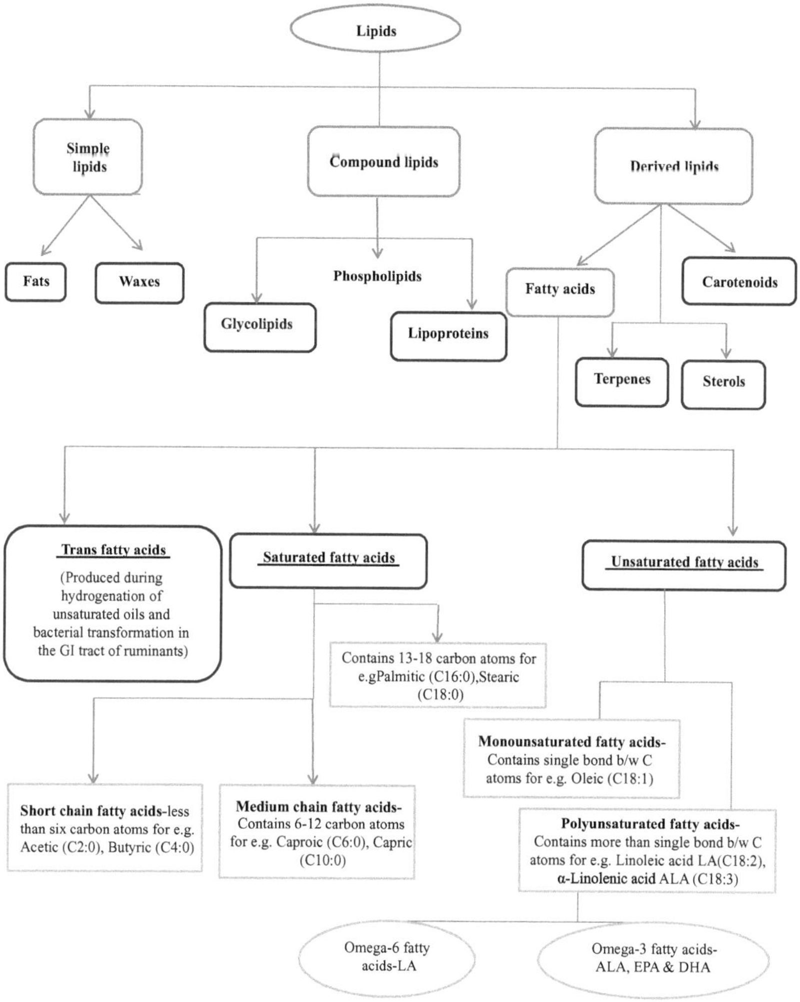

Figure 12.1 Classification of fatty acids

12.2 ESSENTIAL FATTY ACIDS: DIETARY SOURCES AND HEALTH BENEFITS

12.2.1 Omega-9 (ω-9) Fatty Acids

Among all the MUFAs, oleic acid 18:1 (also known as ω-9 fatty acid) is considered as the healthy fat; it is found in olive oil, grapeseed oil, corn oil and sea buckthorn oil. Olive oil is the main dietary source (represents 70–80% of the total olive oil composition) of oleic acid, and it is largely consumed in the diet of Mediterranean region (Owen et al. 2000). In addition to this, many studies have shown health benefits of MUFAs (Table 12.1). MUFAs (mainly oleic acid) have been proved beneficial for the heart as it helps in more secretion of larger chylomicrons that contain greater amount of exogenous dietary triglycerides, which results in faster clearance of postprandial fats and reduces the risk of cardiovascular diseases and hypertension (Bergouignan et al. 2009). Several studies reported that oleic acid reduces the risk of ischemic heart diseases by inhibiting platelet aggregation as well as the secretion of serotonin (Cocchi et al. 2009). Oleic acid is also considered to be highly stable against oxidation, hence preventing chronic heart diseases (Moser et al. 2007). Furthermore, oleic acid was demonstrated to induce positive effects in terms of anticancer activity, which could be explained by its interaction with proteins and an activating factor (nitric oxide) in the body, and it formed a nitric oxide-derived endogenous lipid product which had antioxidant and anti-inflammatory properties (Ruggiero et al. 2014; Liu et al. 2013). Nitric oxide is also essential for skin wound healing, as it stimulates the functions of fibroblasts, keratinocytes and macrophages during the healing process (Frank et al. 2002). One other study showed that oleic acid contributes a part of brain fatty acids and is therefore essential for brain health (Bourre 2009). An oleic acid-rich diet has also been proven to have a cholesterol-lowering effect along with maintaining the HDL (high-density lipoprotein) cholesterol level and reduces triglyceridemia (Priore et al. 2017). Thus, overall the studies suggested that it is important to consume oleic acid in the diet and to replace unhealthy fats with MUFAs.

12.2.2 Omega-6 (ω-6) Fatty Acids

The most common ω-6 fatty acid is linoleic acid 18:2 (LA), which is found naturally in vegetable oils, including sunflower, soybean, safflower, canola and corn oils. LA can be converted into other long-chain ω-6 fatty acids such as arachidonic acid (20:4). LA is also categorized as an EFA because the human body cannot synthesize it. Moreover, the mammalian body cannot convert ω-6 fatty acids to ω-3 fatty acids due to the lack of ω-3 desaturase enzyme (Gogus and Smith 2010). The ω-6 fatty acids are precursors to cannabinoids (alter neurotransmitter release in the brain), lipoxins (bioactive autacoid metabolites of arachidonic acid) and specific eicosanoids (made by the oxidation of PUFAs). These eicosanoids are bioactive compounds;

Table 12.1 Health Benefits of EFAs

EFAs	Structure	Health Benefits	References
Omega-9 fatty acid- oleic acid	CH_3-$(CH_2)_7$-CH=CH-$(CH_2)_7$-COOH	Reduces risk of CHD	Bergouignan et al. (2009)
		Prevents dental caries by inhibition of glucosyltransferase	Choi et al. (2010)
		Acts as anticarcinogenic	Ruggiero et al. (2014)
		Reduces hypertension by inhibiting cholesterol synthesis	Terés et al. (2008)
		Lowers obesity	Yang et al. (2016)
		Improves insulin sensitivity	Sears and Perry (2015)
		Reduces inflammation	Carrillo et al. (2012)
Omega-6 fatty acid- linoleic acid	CH_3-$(CH_2)_4$-CH=CH-CH_2-CH=CH--$(CH_2)_7$-COOH	Acts as anticarcinogenic	Majumdar et al. (2002)
		Prevents atherosclerosis	Sato et al. (2005)
		Maintains lipid metabolism	Pariza et al. (2003)

(*Continued*)

Table 12.1 (Continued) Health Benefits of EFAs

EFAs	Structure	Health Benefits	References
		Helps in bone formation	Watkins et al. (2003)
		Regulates immune response	Albers et al. (2003)
		Regulates vitamin A metabolism	Carta et al. (2014)
		Prevents type II diabetes	Song et al. (2016)
Omega-6 fatty acids- α-linolenic acid (ALA), eicosapentaenoic acid (EPA) and docosahexaenoic acid (DHA)	CH_3-CH_2-CH=CH-CH_2-CH=CH—CH_2-CH=CH-$(CH_2)_7$-COOH (ALA) CH_3-CH_2-CH=CH-CH_2-CH=CH—CH_2-CH=CH- CH_2-CH=CH—CH_2-CH=CH-$(CH_2)_3$-COOH (EPA) CH_3-CH_2-CH=CH-CH_2-CH=CH—CH_2-CH=CH- CH_2-CH=CH—CH_2-CH=CH-CH_2-CH=CH-$(CH_2)_2$-COOH (DHA)	Prevents cardiovascular diseases	Balk et al. (2007)
		Prevents ADHD	Gillies et al. (2012)
		Reduces disease-promoting inflammatory response	Calder (2006)
		Acts as neuroprotection	Blondeau et al. (2015)
		Prevents autoimmune diseases	Itariu and Stulnig (2014)
		Acts as antiarrhythmic	Marchioli et al. (2005)
		Helps in mental development	Guxens et al. (2011)
		Reduces occurrence of Alzheimer's disease	Oster and Pillot (2010)

act as antithrombotic, vasodilating and anti-inflammatory; and play a role in platelet aggregation (Simopoulos 2006a,b).

One of the other geometric and positional isomers of linoleic acid is conjugated linoleic acid (CLA), which also possesses various health benefits. It is obtained naturally from the rumen and the mammary gland during the bio-hydrogenation of linoleic acid to stearic acid by *Butyrivibrio fibrisolvens* bacteria and during the bio-hydrogenation of linoleic or linolenic acid by 9-desaturase enzyme, respectively. Various experimental studies reported the ameliorating effects of CLA on carcinogenesis (Majumdar et al. 2002); it is observed that CLA participates in all three biological events (initiation, promotion and progression) of carcinogenesis, inhibiting the growth and proliferation, inducing apoptosis and diminishing the density of the cancerous cells (Parodi 2001; Belury et al. 2002). In sum, the inhibitory effects of CLA can be characterized as those of antioxidants (Sato et al. 2005). Moreover, CLA helps in maintaining the lipid metabolism (Pariza et al. 2003), bone metabolism (Watkins et al. 2003), immune response (Albers et al. 2003) and vitamin A metabolism (Carta et al. 2014). One recent study showed that feeding fermented milk (containing CLA) for 6 weeks showed a beneficial effect on type II diabetes (Song et al. 2016).

12.2.3 Omega-3 (ω-3) Fatty Acids

These include α-linolenic acid 18:3 (ALA), eicosapentaenoic acid 20:5 (EPA) and docosahexaenoic acid 22:6 (DHA). ALA is abundant in our diet; however, getting sufficient amounts of EPA and DHA through diet alone can be challenging. All these ω-3 fatty acids affect various functions in the body such as membrane composition, gene expression and cell signaling (Shahidi and Miraliakbari 2005). ALA is plentiful in many plant foods, including walnuts, canola, flaxseed, several legumes and green leafy vegetables (Gebauer et al. 2006), while fish oil is the richest source of EPA and DHA. ALA is also the precursor of two important longer-chain fatty acids, namely EPA and DHA, which play vital roles in maintaining the cardiovascular health, brain development and function and also have inflammatory responses (Das 2006; Wang et al. 2006). As we know, fats are absorbed very proficiently from the digestive tract into the body. Similarly, ALA is absorbed 96% or more in the body (Burdge 2006) due to its several possible metabolic fates that occur in the bloodstream. The body can store the ALA in adipose tissue, and further use it to produce acetyl-CoA, produce energy through β-oxidation, synthesize other non-essential saturated or monounsaturated fatty acids or convert it to EPA and DHA (Stark 2008).

In humans, ALA is partially converted to EPA (i.e. 8–20%), whereas the conversion rates of ALA to DHA are 0.5–9% (Burdge 2004; Defilippis and Sperling 2006). This conversion rate is also gender dependent. Several studies reported that women of younger age showed a substantially 2.5-fold greater rate of conversion of ALA to EPA than healthy men. It shows that women have a lower utilization of ALA to β-oxidation and other processes; hence, it is more available for conversion (Burdge 2004). Sex differences have also been observed in the conversion rates of ALA to DHA. In males it is observed that only 0.5–4% of ALA is converted to DHA, whereas in females the conversion rate is 9% (Burdge 2004; Defilippis and Sperling 2006).

Various researchers have reported the protective role of ALA, EPA and DHA in coronary heart disease (CHD) (Fattore and Massa 2018). ALA, as a precursor of EPA and DHA, has been linked with the maintenance of vascular tone, serum lipid level, heart rate, inflammatory responses, blood pressure, arrhythmia and platelet function in the body (Robinson and Stone 2006). The ω-3 fatty acids also can raise the good HDL cholesterol levels in the body while reducing the bad cholesterol. Experimental studies have also been carried out to assess the effects of ω-3 fatty acid–enriched diet in the prevention of inflammatory diseases (Laye et al. 2018). Several diseases such as Crohn's disease, rheumatoid arthritis, chronic obstructive pulmonary disease, psoriasis and irritable bowel syndrome also occur with high levels of inflammation. Therefore, it is observed that ω-3 fatty acids are able to prevent disease-related inflammatory responses. It has been shown that an ALA-enriched diet provokes anti-inflammatory effects by inhibiting IL-1β, IL-6 and TNF-α production in peripheral blood mononuclear cells (Laye et al. 2018). The ω-3 fatty acid DHA is the most abundant fatty acid in the human brain, and its quantity in the brain membrane depends upon the amount of ω-3 fatty acids in the diet and the person's age (decreasing with aging) (Uauy and Dangour 2006). DHA also contributes to the structure of the eye retina and nervous system. Hence, an ample supply of this fatty acid is essential for the proper development of vital organs. Some studies suggested that ω-3 fatty acids have physiological importance in protection from various disorders and diseases such as attention deficit hyperactivity disorder (ADHD), bipolar disorder, depression, dyslexia, schizophrenia and auto-immune disorders. Various studies have also reported that dietary intake of ω-3 fatty acids during pregnancy was associated with fetal development and improved problem-solving skills of the child during infancy (Judge et al. 2007).

Due to the numerous health benefits of ω-3 fatty acids (ALA, EPA, DHA), they are essential to staying fit, developing the mammalian body and sustaining healthy life. They are involved in every cell of the body and also act as structural components of the cell membrane. Since the body cannot synthesize sufficient amounts of these essential nutrients, they must be taken through diet.

12.3 DEVELOPMENT OF EFA-ENRICHED FUNCTIONAL FOODS: CHALLENGES AND OPPORTUNITIES

There are enormous health benefits of consuming long-chain ω-3 and ω-6 fatty acids that are reported by researchers from time to time (Shahidi and Ambigaipalan 2018). Presently, consumers select food with a great deal of awareness of the healthy effects of EFAs. A lacuna in the EFA availability in our diet is prompting researchers cum food technologists to work in this direction by means of fortification, enrichment, supplementation and various other means. Their efforts are depicted in Table 12.2, where we have long lists of such products that help us to stay fit. These changes can be executed without any major modification in our daily diet patterns (Decker et al. 2012). However, a challenge in this modification is to address the EFA-fortified food products' susceptibility to oxidation, which ultimately affects its flavor via degradation and hence ceases consumer preference for these products. The shelf life of formulated products is also diminished. Among the selected efforts in the direction of preventing the oxidation of EFAs to make them shelf stable and more palatable (Anbudhasan et al. 2014), microencapsulation is the best means to protect the EFAs in food during processing and storage (Ye et al. 2009). It possesses a coating of an active substance (core) within the matrix (carrier material, wall material or external phase) (Desai and Park 2005). Fish oil and long-chain unsaturated fatty acid supplements can be encapsulated to prevent oxidation by preventing contact between oxygen, air, light and moisture (Drusch and Berg 2008). Water-soluble and -insoluble microencapsulates used nowadays in food formulations depend upon the nature and chemical composition of the food. Another important requisite of microencapsulation is the bioavailability of EFAs upon consumption or dissociation of microencapsulation in the gastrointestinal tract (Kaur et al. 2015). Based upon these two abovementioned requirements, that is, the nature of food and the bioavailability of EFAs in the human gastrointestinal tract, the below-listed techniques (physical and chemical) can be utilized for efficient microencapsulation.

Table 12.2 Food Fortified with EFAs

Foods	References
Eggs	Bourre and Galea (2006)
Milk	Lopez-Huertas (2010)
Dairy products	Kolanowski and Weißbrodt (2007)
Dairy beverages	Özer and Kirmaci (2010)
Bread	Yep et al. (2002)
Pasta	Verardo et al. (2009)
Granola bars	Nielsen and Jacobsen (2009)
Surimi seafood	Pérez-Mateos et al. (2004)
Infant formulas and baby foods	Bosewell et al. (1996)
Rosemary and green tea extract	Pérez-Mateos et al. (2006)
Pomegranate juice	Habibi et al. (2017)
Biscuit	Umesha et al. (2015)

12.3.1 Physical Techniques

Spray drying, complex coacervation, extrusion, melt injection and fluidized bed drying are among the physical techniques frequently in operation for microencapsulation.

12.3.1.1 *Spray Drying* is the most predominant and inexpensive technology for microencapsulation (Laohasongkram et al. 2011). It works on the principle of atomization of the feed through a pressure nozzle (atomizer) into super-heated steam (commonly air), and it results in the conversion of a liquid form (solution and dispersion) into dry powder. Generally by spray drying, oil-in-water emulsion type matrix structures have been made in which oil droplets (core) are embedded in a continuous phase (water soluble) of carrier material (Jafari et al. 2008; Barrow et al. 2013). The most common water-soluble carrier materials are polysaccharides (modified cellulose, maltodextrin), proteins (soya protein, gelatin, milk protein), gums (pullulan, xanthan, alginate) and lipids (waxes) (Devi et al. 2016; Erminio 2007).

12.3.1.2 *Coacervation* is a phase separation technique that involves the separation of a homogeneous liquid phase of coating material from a polymeric solution and coating that phase as a uniform layer around suspended core particles. When only a single polymer is present, it is referred to as simple coacervation; however, when there are two or more polymers of opposite charges, it is known as complex coacervation (Drusch and Mannino 2009). For encapsulation, complex coacervation is preferred. It is a continuous process which involves several steps: The first step is dispersion of core material into a solution containing coating material. The second step includes deposition of the insoluble polymer solution upon the core material. In the last step, the prepared microcapsules are stabilized by cross-linking, thermal treatment or desolvation (Comunian et al. 2013).

12.3.1.3 *Extrusion* is a mechanical process in which a carbohydrate matrix is used to encapsulate core material. It is commonly used for encapsulation of fish oil and has very good barrier properties. This process is accomplished at low temperatures (<30°C) and low pressures (500–5,000 kPa). This preparation involves protein, gum or modified starch as an emulsifier and mixing with matrix materials (starch, flours, proteins, gums, etc.) further; in the mixing step, plasticizer (glycerol and water) and anti-oxidant (ascorbic acid) are added and passed through a twin-screw extruder.

12.3.1.4 *Melt Injection* also uses a carbohydrate matrix and is generally employed for the encapsulation of fish oil. This technique involves mixing fish oil with antioxidants, water and emulsifiers and then heating at 100°C followed by filtration and collection with cold organic solvent (such as liquid nitrogen or iso-propanol), which solidifies the matrix and alters it into a glassy state material (Valentinotti et al. 2006).

12.3.1.5 *Fluidized Bed Drying* includes sprinkling of coating material (liquid) onto the particles (core material) followed with rapid evaporation in the presence of hot air. Fish oil powder could be coated in the presence of silica (flow aid), hydrogenated palm wax and antioxidant (Ponginebbi and Publisi 2008). Here, application of the process yielded high amounts of encapsulated material; thus it is economic by commercial means.

12.3.2 Chemical Techniques

Chemical techniques include emulsion polymerization, cross-linking, calcium carbonate capsules and alginate-based microspheres.

12.3.2.1 *Emulsion Polymerization* is a technique in which a monomer (coating material) is added dropwise to the aqueous polymer medium containing core material and emulsifier. The type and charge of emulsifier influence the oxidation reaction to a greater extent (Mojgan et al. 2014). Anionic emulsifiers attract positively charged metal ions such as iron, resulting in accelerated oxidation. Therefore, for fish oil, cationic emulsifiers such as proteins are preferred as they prevent oxidation by repelling metals. A protein layer may protect oil by its free radical scavenging properties or by steric hindrance (Kaur et al. 2015).

12.3.2.2 *Cross-Linking* is also a chemical process in which protein and polysaccharide are used as coating material which further holds a core material by cross-linking. This process is employed either at high temperature (above 500°C) or by the use of a cross-linking agent (includes formaldehyde, terephthaloyl chloride, etc). This is a versatile method and can be accepted for microencapsulation of liquid or solid, soluble or insoluble materials, and for the production of both microcapsules and nanocapsules (Gharibzahedi et al. 2018).

12.3.2.3 *Calcium Carbonate Adsorption* is carried out by electrostatically adsorbing calcium carbonate particles onto the oil surface (negatively charged), which forms a fish oil-in-water emulsion (Nakahara et al. 2006). This calcium shell only dissolves at low pH, once the capsules reach to the stomach. The advantage of this technique is in the particle size: the calcium shell is very small (approximately 20 μm), much smaller than that obtained with other techniques. However, the process is expensive to operate (Desai and Park 2005).

12.3.2.4 *Alginate-Based Microspheres* are widely used in the encapsulation of ω-3 fatty acids. Alginates are derived from brown algae and hence are considered as nontoxic and biocompatible. Moreover, alginate-based microspheres can be made at lab scale. It is also prepared by spray drying and fluidized bed drying processes (Drusch and Mannino 2009).

12.4 TAILOR-MADE FATS: TYPES AND APPLICATION

Tailor-made fats (also known as structured lipids) are triacylglycerols (TAGs) that have been made by the incorporation of new fatty acids and restructured to change the positions of fatty acids (Osborn and Akoh 2002). TAGs have been made to achieve desirable properties in the body such as absorption, metabolism and digestion, and they also can be nutraceuticals capable of being used in food with lower energy value to produce numerous health benefits (Zam 2015). Tailor-made fats can be made by chemical and enzymatic means by the processes of direct esterification, inter-esterification, acidolysis and alcoholysis. Basically, these are made by substitution of the glycerol moiety of triglycerides with alternative alcohols such as sugar alcohols and the substitution of long-chain fatty acids with short-, medium- and long-chain fatty acids (inter-esterification) (Zam 2015). Inter-esterification is a chemical reaction which requires quite high temperatures and anhydrous conditions and further produces a mixture of short- and medium-chain fatty acid moieties in the triglyceride backbone. Being inexpensive and easy to scale up, it has been widely adopted in food industries (Willis and Marangoni 1999). Yet, from the viewpoint of producing lipids with very specific functional and medicinal properties, enzymatic synthesis is preferred (Willis et al. 1998). During enzymatic synthesis, *lipase* enzymes have been used to speed up various reactions, viz. direct esterification, alcoholysis and acidolysis. Due to lipase selectivity and regio-specificity, enzymatic reactions offer greater control over the positional distribution of fatty acids in the resulting product (Balcão and Malcata 2002). Various kinds of lipases from plant, microbial and animal sources have been used for enzymatic modification of lipids. Among these, microbial lipases are the most preferred for lipid modification (Xu 2000). Genetic modification is also done to change the fatty acid profile of certain oils which are used for food purposes. The basic methods for genetic modification of fatty acid composition are cloning and gene transference from one plant species into another species to achieve the desired fatty acid profile (Miquel and Browse 1995).

Tailor-made fats possess various health benefits such as prevention of thrombosis, improved immune function, enhanced nitrogen balance, prevention of obesity and reduced cholesterol and risk of cancer (Osbor and Akoh 2002). Another study showed that TAGs are rapidly oxidized and cleared faster from the plasma (Kruimel et al. 2000).

Chemical and enzymatic inter-esterification is widely preferred for the production of tailor-made fats (shortening, spreads and margarine) with zero or minimum trans fat. Several vegetable oils such as corn, cottonseed, palm, sunflower, peanut and canola can be inter-esterified with fully hydrogenated soybean oil or cottonseed fat to develop desirable fatty acid compositions for margarine, spreads and shortenings (Puligundla et al. 2012).

Tailor-made fats have wide applications (Table 12.3) in the food industry due to their desirable physical properties (such as melting point, slow rancidity and composition). These are used as an alternative to cocoa butter for the production of confectionery products. In recent years vegetable oils such as kokum oil, palm oil and tea seed oil have been used to prepare cocoa butter replacers, which have properties very similar to those of cocoa butter (Pinyaphong and Phutrakul 2009). Moreover, genetically modified soybean and sunflower oil have been used for frying and bakery goods without hydrogenation. These oils also have better oxidation stability and fatty acid profiles (Fick 1983). Tailor-made fat is

Table 12.3 Examples of Tailor-Made Fats/Modified Fats

Modified Fat	Composition	Applications	References
Caprenin	Esterified glycerol with caprylic (C8:0), capric C10:0) and behenic (C22:0) fatty acids	Used in chocolate bars, soft candies, and confectionery coatings for nuts, fruits and cookies	Sandrou and Arvanitoyannis (2000)
Olestra	Acylated sucrose polyester	Used in ice cream, margarine, cheese and baked goods, and can be blended with vegetable oil	Hunt et al. (1998)
Salatrim/Benefat	Structured triglycerides comprised of a mixture of short- and long-chain fatty acids	Used in chocolate coatings, caramels, toffees, fillings for confectionery and baked goods, peanut spreads, savory dressings, dips, sauces and dairy products	Zam (2015)
Neobee	Capric and caprylic acids	Used as a flavor carrier, textural component and replacement for butter oil	Zam (2015)
Betapol	Enzymatically synthesized	Used in infant formulas	Osborn and Akoh (2002)
Emulsifiers	Sucrose fatty acid esters, lecithin, mono- and diglycerides, lecithin, sodium stearoyl-2-lactylate and polyglycerol esters	Used in margarines, frozen desserts, baked goods, dairy products, spreads, shortenings, processed meats, cake frostings whipped toppings and confections	Osborn and Akoh (2002)

(*Continued*)

Table 12.3 (Continued) Examples of Tailor-Made Fats/Modified Fats

Modified Fat	Composition	Applications	References
Sorbestrin	Mixture of esters of sorbitol and sorbitol anhydrides with fatty acids	Used in salad dressings, baked goods and frying	Osborn and Akoh (2002)
Dialkyldihexadecylmalonate (DDM)	Fatty alcohol dicarboxylic acid, ester of malonic acid and alkylmalonic acid	Used for frying	Artz et al. (2005)
Esterified propoxylated glycerols (EPGs)	Derivatives of propylene oxide	Used in frozen desserts, salad dressings, spreads, baked goods, and for cooking and frying	Ognean et al. (2006)
Trialkoxytricarballylate (TATCA), trialkoxycitrate (TAC) and trialkoxyglyceryl ether (TGE)	Polycarboxylic acids with carboxylic acid groups esterified with fatty acids	Used for frying	Akoh (1998)

also utilized for the development of human milk fat substitutes by inter-esterification; the substitutes have the same fatty acid profile as human milk (Yang et al. 2003). Various modified fats available in the market are listed in Table 12.3.

12.4.1 Caprenin

Caprenin (common name: Caprocaprylobehenin) a modified fat made by esterification of glycerol with caprylic, capric and behenic fatty acids isolated from coconut, rapeseed oils and palm kernel (Costin and Segal 1999). The FDA identified it as GRAS (Generally Recognized as Safe), it is readily metabolized in the body and it provides only 5 kcal/g (Akoh 1998). Caprenin has properties similar to those of cocoa butter; hence it is used in candy bars and in confectionery coatings for nuts and cookies. Whenever it is utilized along with polydextrose, it proves itself best, and hence commercially labeled reduced-calorie and reduced-fat chocolate bars can be made with it (Sandrou and Arvanitoyannis 2000). The data obtained by Swift et al. (1992) revealed that whenever caprenin was given to male subjects as part of a low-fat diet for 6 days, plasma cholesterol concentration didn't vary; however, HDL cholesterol concentration was reduced by 14%. Similarly, a medium-chain triacylglycerol diet elevates plasma triglycerides by 42% and HDL cholesterol reduces by 15%.

12.4.2 Olestra

Olestra (esterified sucrose) is made by trans-esterification or inter-esterification of sucrose with long-chain fatty acids isolated from vegetable oils (Peters et al. 1997). Olestra can be solid and liquid at room temperature depending on the composition of the fatty acids from which it is derived (Harrigan and Breene 1989). Olestra constitutes a large amount of non-polar fatty acids which cannot be hydrolyzed by gastric and pancreatic enzymes. Olestra also remains intact when it passes through the gastrointestinal tract and is not absorbed in the body; hence, it provides zero calories to the diet (Cooper et al. 1997, Haumann 1986). The FDA has approved the use of olestra as a replacement for fats and oils. Therefore, it is used in the development of various food products such as ice cream, cheese, margarine and baked goods. The olestra fatty acid profile also makes it thermostable, so it can be used for frying (Akoh 1998; Cooper et al. 1997; Peters et al. 1997).

12.4.3 Salatrim

Salatrim (also known as Benefat) is a structured triglyceride comprised of a mixture containing at least one short-chain fatty acid (predominantly C2:0, C3:0 or C4:0 fatty acids) and at least one long-chain fatty acid (primarily C18:0 fatty acid) randomly attached to a glycerol backbone (Kosmark 1996). Salatrim is made by base-catalyzed inter-esterification of hydrogenated vegetable oils with TAGs of acetic, propionic or butyric fatty acids (Smith et al. 1994). Salatrim behaves in the same manner as Caprenin; it is a calorie-poor fat that provides 5 kcal/g. The stearic acid in Salatrim is only 50% absorbed in the body, while acetyl and propionyl acid are readily hydrolyzed by lipases in the stomach and upper intestine and readily converted to carbon dioxide. It is also GRAS and allowed to be used in many bakery, confectionery and dairy blends (such as frozen dessert, sour cream and cheese). However, Salatrim is not suitable for frying due to its thermo-instability (Hayes et al. 1994; Klemann et al. 1994).

12.4.4 Neobee

Neobee is another fat replacer which is comprised of capric and caprylic fatty acid. Commercially, these are listed as Neobee 1053, Neobee 1814, Neobee M-5 (made up of both capric and caprylic acids) and Neobee 1095 (contains only capric acid) based on the difference in their physical and chemical properties (Heydinger and Nakhasi 1996). Neobee 1095 is solid in nature thus not suited for the formulation of all food products. Neobee 1814 is a medium-chain triglyceride produced by inter-esterification of fatty acid with butter oil and it is appropriate to replace butter oil in a range of products. It also serves as a flavor carrier and functions as a textural component for low-fat products (Babayan et al. 1997).

12.4.5 Betapol

Betapol is recognized as the first commercially available enzymatically synthesized modified fat. It is produced by the reaction between tripalmitin and UFAs in the presence of lipase. Betapol was designed to imitate the fatty acid profile of human milk for use in infant formulas (Osborn and Akoh 2002).

Much more tailoring has been done in fats that are still not commercially available in the common market. These include sorbestrin, dialkyldihexadecylmalonate (DDM), esterified propoxylatedglycerols (EPGs), trialkoxytricarballylate (TATCA), trialkoxy citrate (TAC) and tri alkoxyglyceryl ether (TGE) (Ognean et al. 2006).

12.5 TOXICITY AND SAFETY ISSUES

It is common for users and consumers to be concerned about the intake and health consequences directly associated with saturated and trans fatty acids (Table 12.4). The intake of TFAs elevates the level of low-density lipoprotein (LDL) cholesterol, insulin and triglycerides and decreases the level of beneficial high-density lipoprotein (HDL) cholesterol. Based on some epidemiological studies, it has been proven that these lead to changes in blood composition and escalate the risk of cardiovascular diseases. The major dietary sources of TFAs are products formulated with hydrogenated fats,

Table 12.4 Adverse Health Effects of Fatty Acids

Fatty Acids	Adverse Health Effects	References
Saturated and trans fatty acids	Cardiovascular diseases	Mozaffarian et al (2006); Iqbal (2014)
	Interference with EFAs	Dhaka et al. (2011)
	Breast cancer	Chajès et al. (2008); Bougnoux et al. (2010)
	Obesity	Thompson et al. (2011)
	Diabetes	Thompson et al. (2011)
	Colon cancer	Michels et al. (2018)
	Allergy (asthma and allergic cold)	Wendell et al. (2014)
Monounsaturated fatty acids	Acute respiratory distress syndrome (ARDS)-lung injury	Gonçalves-de-Albuquerque et al. (2015)
Polyunsaturated fatty acids	Increased incidence of prostate cancer risk	Khankar et al. (2016)
	Mild gastrointestinal symptoms	Balk et al. (2007)
	Oxidative and inflammatory stresses	Sato et al. (2005)

which include margarines, bakery products, shortenings and fast foods as mentioned above. Hence, intake of significant levels of TFAs has been a major health concern for consumers and regulatory agencies. Therefore, food technologists are increasingly focusing on the development of food products with zero or minimum trans fat. Several techniques have been used to reduce or minimize trans fatty acid in processes such as food reformulations, substitution of saturated and trans fat with plant oil (Mozaffarian et al. 2010), fat inter-esterification (Baumgartner and Mensink 2008) and the use of genetically modified fatty acids (Liu et al. 2011). Various studies also reported the direct link of TFAs with breast and colon cancer and cardiovascular diseases, and TFAs also reduce the length of pregnancy, affect the nervous system, increases the occurrence of preeclampsia, impairs vision in infants and also results in obesity, diabetes and allergy.

Despite the proven health benefits of long-chain polyunsaturated fatty acid for humans, paradoxically, more recent studies suggested detrimental effects of high consumption of PUFAs (Table 12.4) (Berry 2001). Studies showed that high consumption of ω-6 fatty acids exaggerates hyper-insulinemia, insulin resistance, atherosclerosis and carcinogenesis (Okuyama et al. 1997). However, these diseases are multi-factorial in origin; therefore lifestyle and environmental factors (such as exercise, smoking and pollution) must be considered in relation to dietary patterns (Lehnen et al. 2015). There are several ways by which ω-6 fatty acids interfere with the metabolism of ALA, EPA and DHA, thus adversely affecting the normal development of the brain, retina and blood clotting mechanism in the body (Berry 2001). Excess LA increases the ratio of ω-6 to ω-3 fatty acids and hence affects numerous pathophysiological processes, which range from inflammatory reactions to atherogenesis and carcinogenesis (Okuyama et al. 1997). Moreover, ω-6 and ω-3 fatty acids compete for the same enzymes which participate in the elongation and desaturation of their respective pathways. However, the desirable level of ω-6 and ω-3 fatty acids in the diet is still unknown.

In the case of ω-3 fatty acids the most common adverse effects are gastrointestinal disturbance, nausea and fishy burp (after consumption of fish). Moreover, one experimental study also reported an increased incidence of prostate cancer risk in men with high consumption or high blood levels of ALA (Brouwer et al. 2004). Therefore, accurate proportions of ω-6 and ω-3 fatty acids should be included in the diet. As per the Dietary Reference Intake, ALA should provide 0.6–1.2% of the energy in the body with up to 10% provided by PUFAs. The FDA also recommended a 3 g/day dose of ω-3 fatty acids in the diet (Roosha and Parloop 2010).

12.6 CONCLUSION

Based on previous research and studies, it can be concluded that PUFAs are indispensable for maintenance of healthy life. Therefore, it is important to take an appropriate amount of PUFAs in the diet. In recent years various techniques such as microencapsulation have been used to develop EFAs enriched functional food. However, due to the oxidative instability of fatty acids, these products have poor shelf life. Therefore, it is important to conduct further research that will allow more control over the oxidation reaction.

In modern days, consumer concerns about overweight, cancer and cardiovascular diseases have forced food scientists to develop modified fats and fat substitutes that have similar functional, physical and sensory properties to those of traditional fat, besides its nutritional and medicinal properties. Food technologists have also developed a number of modified fats using several technologies (such as chemical, enzymatic and genetic modification). Research on modified fat is an interesting area that provides opportunities for entrepreneurs to develop functional foods with extended shelf life. However, further research is needed to reduce the cost and to maintain controlled conditions during scale-up.

REFERENCES

Akoh, C. C. (1998). Fat replacers. *Food Technology*, 52, 47–52.

Albers, R., van der Wielen, R. P., Brink, E. J., Hendricks, H. F., Taran-Dorovska, V. N., & Mohede, I. C. (2003). Effects of cis-9, trans-11 and trans-10, cis-12 conjugated linoleic acid (CLA) isomers on immune function in healthy men. *European Journal of Clinical Nutrition*, 57(4), 595–603.

Anbudhasan, P., Surendraraj, A., Karkuzhali, S., & Ramasamy, D. (2014). Development of omega 3 fatty acid enriched stable functional foods: Challenges and options. *International Journal of Innovative Research and Review*, 2(1), 1–13.

Artz, W. E., Osidacz, P. C., & Coscione, A. R. (2005). Iron accumulation in oil during the deep-fat frying of meat. *Journal of the American Oil Chemists' Society*, 82, 249–254.

Babayan, V. K., Blackburn, G. L., & Bistrian, B. R. Structured lipid containing dairy fat. US Patent; 1997. No. 4,952,606.

Balcão, V. M., & Malcata, F. X. (2002). Enzyme-mediated modification of milk fat. In: Kuo, G. T. M., editor. *Lipid Biotechnology*, Marcel Dekker: New York, pp. 479–492.

Balk, E. M., Horsley, T. A., Newberry, S. J., et al (2007). A collaborative effort to apply the evidence-based review process to the field of nutrition: Challenges, benefits, and lessons learned. *American Journal of Clinical Nutrition*, 85(6), 1448–1456.

Barrow, C. J., Wang, Bo, Adhikari, Benu, & Liu, H. (2013). *Spray Drying and Encapsulation of omega-3 Oils*. doi:10.1533/9780857098863.2.194.

Baumgartner, S., & Mensink, R. P. (2008). Metabolic effects of interesterified dietary fat in humans: A literature review.

Belury, M. A., Moya-Camarena, S. Y., Lu, M., Shi, L., Leesnitzer, L. M., & Blanchard, S. V. (2002). Conjugated linoleic acid is an activator and ligand for peroxisome proliferator-activated receptor-gamma (PPARγ). *Nutrition Research*, 22(7), 817–824.

Bergouignan, A., Momken, I., Schoeller, D. A., Simon, C., & Blanc, S. (2009). Metabolic fate of saturated and monounsaturated dietary fats: The Mediterranean diet revisited from epidemiological evidence to cellular mechanisms. *Progress in Lipid Research*, 48(3–4), 128–147.

Berry, E. M. (2001). Are diets high in omega-6 polyunsaturated fatty acids unhealthy? *European Heart Journal Supplements*, 3(suppl_D), D37–D41.

Blondeau, N., Lipsky, R. H., Bourourou, M., Duncan, M. W., Gorelick, P. B., & Marini, A. M. (2015). Alpha-linolenic acid: An omega-3 fatty acid with neuroprotective properties—Ready for use in the stroke clinic? *BioMed Research International*, 2015, Article ID 519830.

Bosewell, K., Koskelo, E. K., Carl, L., Glaza, S., Hensen, D. J., Williams, K. D., & Kyle, D. J. (1996). Preclinical evaluation of single-cell oils that are highly enriched with arachidonic acid and docosahexaenoic acid. *Food and Chemical Toxicology*, 34(7), 585–593.

Bougnoux, P., Hajjaji, N., Maheo, K., Couet, C., & Chevalier, S. (2010). Fatty acids and breast cancer: Sensitization to treatments and prevention of metastatic re-growth. *Progress in Lipid Research*, 49(1), 76–86.

Bourre, J. M. (2009). Diet, brain lipids, and brain functions: Polyunsaturated fatty acids, mainly omega-3 fatty acids. In: *Handbook of Neurochemistry and Molecular Neurobiology*. New York, NY: Springer, pp. 409–441.

Bourre, J. M., & Galea, F. (2006). An important source of omega-3 fatty acids, vitamins D and E, carotenoids, iodine and selenium: A new natural multi-enriched egg. *Journal of Nutrition, Health and Aging*, 10(5), 371.

Brouwer, I. A., Katan, M. B., & Zock, P. L. (2004). Dietary α-linolenic acid is associated with reduced risk of fatal coronary heart disease, but increased prostate cancer risk: A meta-analysis. *The Journal of Nutrition*, 134(4), 919–922.

Burdge, G. C. (2004). Alpha-linolenic acid metabolism in men and women: Nutritional and biological implications. *Current Opinion in Clinical Nutrition and Metabolic Care*, 7(2), 137–144.

Burdge, G. C. (2006). Metabolism of α-linolenic acid in humans. *Prostaglandins, Leukotrienes and Essential Fatty Acids*, 75(3), 161–168.

Calder, P. C. (2006). n− 3 polyunsaturated fatty acids, inflammation, and inflammatory diseases. *The American Journal of Clinical Nutrition*, 83(6), 1505S–1519S.

Carrillo Pérez, C., CaviaCamarero, M. D. M., & Alonso de la Torre, S. (2012). Role of oleic acid in immune system; mechanism of action; a review. *Nutriciónhospitalaria*, 27(4), 978–990.

Carta, G., Murru, E., Cordeddu, L., Ortiz, B., Giordano, E., Belury, M., Quadro, L., & Banni, S. (2014). Metabolic interactions between vitamin A and conjugated linoleic acid. *Nutrients*, 6(3), 1262–1272.

Chajès, V., Thiébaut, A. C., Rotival, M. et al. (2008). Association between serum trans-monounsaturated fatty acids and breast cancer risk in the E3N-EPIC Study. *American Journal of Epidemiology*, 167(11), 1312–1320.

Choi, S.-G., Won, S.-R., & Rhee, H.-I. (2010). Oleic acid and inhibition of glucosyltransferase. olives and olive oil in health and disease. *Prevention*, 1375–1383. doi:10.1016/B978-0-12-374420-3.00153-4.

Christie, W. W., & Han, X. (2003). *Lipid Analysis: Isolation, Separation, Identification and Lipidomeic Analysis*. The Oily Press.

Cocchi, M., Tonello, L., Martínez, Á. J., Lercker, G., & Caramia, G. M. (2009). Extra virgin olive oil and oleic acid. *Nutrición Clínica y Dietética Hospitalaria*, 29(3), 12–24.

Comunian, T. A., Thomazini, M., Alves, A. J. G., de Matos Junior, F. E., de CarvalhoBalieiro, J. C., & Favaro-Trindade, C. S. (2013). Microencapsulation of ascorbic acid by complex coacervation: Protection and controlled release. *Food Research International*, 52(1), 373–379.

Cooper, D. A., Berry, D. A., Jones, M. B., Kiorpes, A. L., & Peters, J. C. (1997). Olestra's effect on the status of vitamins A, D and E in the pig can be offset by increasing dietary levels of these vitamins. *Journal of Nutrition*, 127, 1589S–1608S.

Costin, G. M., & Segal, R. (1999). *Alimentefuncţionale*, Ed. Academica: Galaţi, pp. 188–202.

Das, U. N. (2006). Essential fatty acids: A review. *Current Pharmaceutical Biotechnology*, 7(6), 467–482.

Decker, E. A., Akoh, C. C., & Wilkes, R. S. (2012). Incorporation of (n-3) fatty acids in foods: Challenges and opportunities. *The Journal of Nutrition*, 142(3), 610S–613S.

DeFilippis, A. P., & Sperling, L. S. (2006). Understanding omega-3s. *American Heart Journal*, 151(3), 564–570.

Denis, I., Potier, B., Heberden, C., & Vancassel, S. (2015). Omega-3 polyunsaturated fatty acids and brain aging. *Current Opinion in Clinical Nutrition and Metabolic Care*, 18(2), 139–146.

Desai, K. G. H., & Park, H. J. (2005). Recent developments in microencapsulation of food ingredients. *Drying Technology*, 23(7), 1361–1394.

Devi, N., Sarmah, M., Khatun, B., & Maji, K. (2016). Encapsulation of active ingredients in polysaccharide-protein complex coacervates. *Advances in Colloid and Interface Science*, 239. doi:10.1016/j.cis.2016.05.009.

Dhaka, V., Gulia, N., Ahlawat, K. S., & Khatkar, B. S. (2011). Trans fats: Sources, health risks and alternative approach: A review. *Journal of Food Science and Technology*, 48(5), 534–541.

Drusch, S., & Berg, S. (2008). Extractable oil in microcapsules prepared by spraydrying: Localisation, determination and impact on oxidative stability determination and impact on oxidative stability. *Food Chemistry*, 109(1), 17–24.
Drusch, S., & Mannino, S. (2009). Patent based review on industrial approaches for the microencapsulation of oils rich in polyunsaturated fatty acids. *Trends in Food Science and Technology*, 20(6), 237–244.
Dyall, S. C. (2011). Methodological issues and inconsistencies in the field of omega-3 fatty acids research. *Prostaglandins, Leukotrienes and Essential Fatty Acids (PLEFA)*, 85(5), 281–285.
Erminio, M. (2007). Use of natural polysaccharides in the microencapsulation Techniques. *Journal of Applied Ichthyology*, 14, 245–249. doi:10.1111/j.1439-0426.1998.tb00650.x.
Fattore, E., & Massa, E. (2018). Dietary fats and cardiovascular health: A summary of the scientific evidence and current debate. *International Journal of Food Sciences and Nutrition*, 69(8), 916–927. doi:10.1080/09637486.2018.1455813.
Fick, G. N. (1983). Genetics and breeding of sunflower. *Journal of the American Oil Chemists' Society*, 60(7), 1252–1253.
Frank, S., Kampfer, H., Wetzler, C., & Pfeilschifter, J. (2002). Nitric oxide drives skin repair: Novel functions of an established mediator. *Kidney International*, 61(3), 882–888.
Gebauer, S. K., Psota, T. L., Harris, W. S., & Kris-Etherton, P. M. (2006). n− 3 fatty acid dietary recommendations and food sources to achieve essentiality and cardiovascular benefits. *The American Journal of Clinical Nutrition*, 83(6), 1526S–1535S.
Gharibzahedi, S. M. T., George, S., Greiner, R., Estevinho, B. N., FrutosFernández, M. J., McClements, D. J., & Roohinejad, S. (2018). New trends in the microencapsulation of functional fatty acid-rich oils using TransglutaminaseCatalyzed crosslinking. *Comprehensive Reviews in Food Science and Food Safety*, 17(2), 274–289.
Gillies, D., Sinn, J. K. H., Lad, S. S., Leach, M. J., & Ross, M. J. (2012). Polyunsaturated fatty acids (PUFA) for attention deficit hyperactivity disorder (ADHD) in children and adolescents (review) 7. The Cochrane Library.
Gogus, U., & Smith, C. (2010). n-3 Omega fatty acids: A review of current knowledge. *International Journal of Food Science and Technology*, 45(3), 417–436.
Gonçalves-de-Albuquerque, C. F., Medeiros-de-Moraes, I. M., de Jesus Oliveira, F. M., Burth, P., Bozza, P. T., Faria, M. V. C., Silva, A. R., & de Castro-Faria-Neto, H. C. (2016). Omega-9 oleic acid induces fatty acid oxidation and decreases organ dysfunction and mortality in experimental sepsis. *PLoS One*, 11(4), e0153607.
Gonçalves-de-Albuquerque, C. F., Silva, A. R., Burth, P., Castro-Faria, M. V., & Castro-Faria-Neto, H. C. (2015). Acute respiratory distress syndrome: Role of oleic acid-triggered lung injury and inflammation. *Mediators of Inflammation*, 2015, 1–9.
Guxens, M., Mendez, M. A., Moltó-Puigmartí, C. et al. (2011). Breastfeeding, long-chain polyunsaturated fatty acids in colostrum, and infant mental development. *Pediatrics-English Edition*, 128(4), e880.

Habibi, A., Keramat, J., Hojjatoleslamy, M., & Tamjidi, F. (2017). Preparation of fish oil microcapsules by complex coacervation of gelatin–gum arabic and their utilization for fortification of pomegranate juice. *Journal of Food Process Engineering*, 40(2), e12385.

Harrigan, K. A., & Breene, W. M. (1989). Fat substitutes: Sucrose esters and simplesse. *Cereal Foods World*, 34, 261–267.

Haumann, B. F. (1986). Getting the fat out – Researchers seek substitutes for full-fat fat. *Journal of American Oil Chemistry Society*, 63, 278–286.

Hayes, J. R., Finley, J. W., & Leveille, G. A. (1994). In vivo metabolism of Salatrim fats in the rat. *Journal of Agriculture and Food Chemistry*, 42(2), 500.

Heydinger, J. A., & Nakhasi, D. K. (1996). Medium chain triacylglycerols. *Journal of Food Lipids*, 3(4), 251–257.

Hunt, Richard, Zorich, L., Nora, & Thomson, A. B. R. (1998). Overview of olestra: A new fat substitute. *Canadian Journal of Gastroenterology = Journal canadien de gastroenterologie*, 12(3), 193–197. doi:10.1155/1998/389685.

Iqbal, M. P. (2014). Trans fatty acids–A risk factor for cardiovascular disease. *Pakistan Journal of Medical Sciences*, 30(1), 194.

Itariu, B. K., & Stulnig, T. M. (2014). Autoimmune aspects of type 2 diabetes mellitus-a mini-review. *Gerontology*, 60(3), 189–196.

Jafari, S. M., Assaidpoor, E., Bhandari, B., & He, Y. H. (2008). Nano-particle encapsulation of fish oil by spray drying. *Food Research International*, 41(2), 172–183.

Judge, M. P., Harel, O., & Lammi-Keefe, C. J. (2007). Maternal consumption of a docosahexaenoic acid–containing functional food during pregnancy: Benefit for infant performance on problem-solving but not on recognition memory tasks at age 9 mo. *The American Journal of Clinical Nutrition*, 85(6), 1572–1577.

Kaur, M., Basu, S., & Shivhare, U. S. (2015). Omega-3 fatty acids: Nutritional aspects, sources, and encapsulation strategies for food fortification. *Direct Research Journal of Health and Pharmacology (DRJHP)*, 3(1), 12–31.

Khankari, N. K., Murff, H. J., Zeng, C. et al. (2016). Polyunsaturated fatty acids and prostate cancer risk: A Mendelian randomisation analysis from the PRACTICAL consortium. *British Journal of Cancer*, 115(5), 624.

Klemann, L. P., Finley, J. W., & Leveille, G. A. (1994). Estimation of the absorption coefficient of stearic acid in Salatrim fats. *Journal of Agriculture and Food Chemistry*, 42(2), 484.

Knauf, V. C., & Del Vecchio, A. J. (1998). Genetic engineering of crops that produce vegetable oil. In: Akoh CC and Min DB, editors. *Food Lipids Chemistry, Nutrition, and Biotechnology*. Marcel Dekker: New York, pp. 779–805.

Kolanowski, W., & Weißbrodt, J. (2007). Sensory quality of dairy products fortified with fish oil. *International Dairy Journal*, 17(10), 1248–1253.

Kosmark, R. (1996). Salatrim: Properties and applications. *Food Technology*, 50, 98–101.

Kruimel, J. W., Naber, A. H., Curfs, J. H., Wenker, M. A., & Jansen, J. B. (2000). With medium-chain triglycerides, higher and faster oxygen radical production by stimulated polymorphonuclear leukocytes occurs. *Journal of Parenteral and Enteral Nutrition*, 24(2), 107–112.

Laohasongkram, K., Mahamaktudsanee, T., & Chaiwanichsiri, S. (2011). Microencapsulation of Macadamia oil by spray drying. *Procedia Food Science*, 1, 1660–1665.
Laye, S., Nadjar, A., Joffre, C., & Bazinet, R. P. (2018). Anti-inflammatory effects of omega-3 fatty acids in the brain: Physiological mechanisms and relevance to pharmacology. *Pharmacological Reviews*, 70(1), 12–38.
Lehnen, T. E., da Silva, M. R., Camacho, A., Marcadenti, A., & Lehnen, A. M. (2015). A review on effects of conjugated linoleic fatty acid (CLA) upon body composition and energetic metabolism. *Journal of the International Society of Sports Nutrition*, 12, 36. doi:10.1186/s12970-015-0097-4.
Liu, S., Jia, Z., Zhou, L. et al. (2013). Nitro-oleic acid protects against adriamycin-induced nephropathy in mice. *American Journal of Physiology-Renal Physiology*, 305(11), F1533–F1541.
Liu, X., Sheng, J., & Curtiss III, R. (2011). Fatty acid production in genetically modified cyanobacteria, 108(17), 6899–6904.
Lopez-Huertas, E. (2010). Health effects of oleic acid and long chain omega-3 fatty acids (EPA and DHA) enriched milks. A review of intervention studies. *Pharmacological Research*, 61(3), 200–207.
Majumdar, B., Wahle, K. W. J., Moir, S., Schofield, A., Choe, S. N., Farquharson, A., Grant, I., & Heys, S. D. (2002). Conjugated linoleic acids (CLAs) regulate the expression of key apoptotic genes in human breast cancer cells. *FASEB Journal*. doi:10.1096/fj.01- 0720fje.
Marchioli, R., Levantesi, G., Macchia, A. et al. (2005). Antiarrhythmic mechanisms of n-3 PUFA and the results of the GISSI-Prevenzione trial. *The Journal of Membrane Biology*, 206(2), 117–128.
Michels, N., Van der Meulen, K., & Huybrechts, I. (2018). Dietary trans fatty acid intake in relation to cancer risk: A systematic review.
Miquel, M., & Browse, J. (1995). Molecular biology of oilseed modification. *Inform*, 6, 108–111.
Mojgan, M., Mahtabani, Amirhossein, & Lotfalieic, Mehrnaz (2014). Emulsion polymerization and encapsulation of micro and nanoparticles within polymer droplets. *Jordan Journal of Chemistry*, 9(4), 229–266. doi:10.12816/0025977.
Morenga, L., & Montez, J. M. (2017). Health effects of saturated and trans-fatty acid intake in children and adolescents: Systematic review and meta-analysis. *PLoS One*, 12(11), e0186672.
Moser, B. R., Sharma, B. K., Doll, K. M., & Erhan, S. Z. (2007). Diesters from oleic acid: Synthesis, low temperature properties, and oxidation stability. *Journal of the American Oil Chemists' Society*, 84(7), 675–680.
Mozaffarian, D., Jacobson, M. F., & Greenstein, J. S. (2010). Food reformulations to reduce trans fatty acids. *New England Journal of Medicine*, 362(21), 2037–2039.
Mozaffarian, D., Katan, M. B., Ascherio, A., Stampfer, M. J., & Willett, W. C. (2006). Trans fatty acids and cardiovascular disease. *New England Journal of Medicine*, 354(15), 1601–1613.
Nakahara, M., Kurosaki, A., Otake, H., Kanai, H., Yokokawa, H., & Kumabe, K. (2006). Method for producing calcium component powder containing oil-soluble substance EP1736532A1.

Nielsen, N. S., & Jacobsen, C. (2009). Methods for reducing lipid oxidation in fish-oil-enriched energy bars. *International Journal of Food Science and Technology*, 44(8), 1536–1546.

Ognean, C. F., Darie, N., & Ognean, M. (2006). Fat replacers. *Journal of Agroalimentary Processes and Technologies*, 12(2), 433–442.

Okuyama, H., Kobayashi, T., & Watanabe, S. (1997). Dietary fatty acids - The n-6/n-3 balance and chronic elderly disease. Excess linolenic acid and relative n-3 deficiency syndrome seen in Japan. *Progress in Lipid Research*, 35, 409–457.

Osborn, H. T., & Akoh, C. C. (2002). Structured lipids-novel fats with medical, nutraceutical, and food applications. *Comprehensive Reviews in Food Science and Food Safety*, 1(3), 110–120.

Oster, T., & Pillot, T. (2010). Docosahexaenoic acid and synaptic protection in Alzheimer's disease mice. *Biochimica et Biophysica Acta (BBA)-Molecular and Cell Biology of Lipids*, 1801(8), 791–798.

Owen, R. W., Mier, W., Giacosa, A., Hull, W. E., Spiegelhalder, B., & Bartsch, H. (2000). Phenolic compounds and squalene in olive oils: The concentration and antioxidant potential of total phenols, simple phenols, secoiridoids, lignans and squalene. *Food and Chemical Toxicology*, 38(8), 647–659.

Özer, B. H., & Kirmaci, H. A. (2010). Functional milks and dairy beverages. *International Journal of Dairy Technology*, 63(1), 1–15.

Pariza, M. W., Park, Y., Xu, X., Ntambi, J., & Kang, K. (2003). Speculation on the mechanisms of action of conjugated linoleic acid. In: J. Sebedio, W. W. Christie and R. Adolf, editors. *Advances in Conjugated Linoleic Acid Research*, Vol. 2. AOCS Press: Champaign, IL, pp. 251–266.

Parodi, P. (2001). Cows milk components with anti-cancer potential. *Australian Journal of Dairy Technology*, 56, 65–73.

Pérez-Mateos, M., Boyd, L., & Lanier, T. (2004). Stability of omega-3 fatty acids in fortified surimi seafoods during chilled storage. *Journal of Agricultural and Food Chemistry*, 52(26), 7944–7949.

Pérez-Mateos, M., Lanier, T. C., & Boyd, L. C. (2006). Effects of rosemary and green tea extracts on frozen surimi gels fortified with omega-3 fatty acids. *Journal of the Science of Food and Agriculture*, 86(4), 558–567.

Peters, J. C., Lawson, K. D., Middleton, S. J., & Triebwasser, K. C. (1997). Assessment of the nutritional effects of olestra, a nonabsorbed fat replacement: Introduction and overview. *Journal of Nutrition*, 127(8), 1539S–1546S.

Pinyaphong, P., & Phutrakul, S. (2009). Synthesis of cocoa butter equivalent from palm oil by *Carica papaya* lipase-catalyzed interesterification. *Chiang Mai Journal of Science*, 36(3), 359–368.

Ponginebbi, L., & Publisi, C. (2008). Coating for oxygen sensitive materials EP1920633 & US2008112987.

Priore, P., Gnoni, A., Natali, F., Testini, M., Gnoni, G. V., Siculella, L., & Damiano, F. (2017). Oleic acid and hydroxytyrosol inhibit cholesterol and fatty acid synthesis in C6 glioma cells. *Oxidative Medicine and Cellular Longevity*, 2017, 1–10.

Puligundla, P., Variyar, P. S., Ko, S., & Obulam, V. S. R. (2012). Emerging trends in modification of dietary oils and fats, and health implications: A review. *SainsMalaysiana*, 41(7), 871–877.

Punia, S., Sandhu, K. S., Siroha, A. K., & Dhull, S. B. (2019). Omega 3-Metabolism, Absorption, bioavailability and health benefits: A review. *PharmaNutrition*, 10(8), 100162.

Robinson, J. G., & Stone, N. J. (2006). Antiatherosclerotic and antithrombotic effects of omega-3 fatty acids. *American Journal of Cardiology*, 98(4A), 39i–49i.

Roosha, P., & Parloop, B. (2010). Omega-3 polyunsaturated fatty acid and cardiovascular disease: A review. *Gujarat Medical Journal*, 65(2), 66–70.

Ruggiero, M., Ward, E., Smith, R. et al. (2014). Oleic acid, deglycosylated vitamin D-binding protein, nitric oxide: A molecular triad made lethal to cancer. *Anticancer Research*, 34(7), 3569–3578.

Sandrou, D. K., & Arvanitoyannis, I. S. (2000). Low-fat/calorie foods: Current state and perspective. *Critical Reviews in Food Science and Nutrition*, 40(5), 427–447.

Sato, M., Shibata, K., Nomura, R., Kawamoto, D., Nagamine, R., & Imaizumi, K. (2005). Linoleic acid-rich fats reduce atherosclerosis development beyond its oxidative and inflammatory stress-increasing effect in apolipoprotein E-deficient mice in comparison with saturated fatty acid-rich fats. *British Journal of Nutrition*, 94(6), 896–901.

Sears, B., & Perry, M. (2015).The role of fatty acids in insulin resistance. *Lipids in Health and Disease*, 14(1), 121.

Shahidi, F., & Miraliakbari, H. (2005). Omega-3 fatty acids in health and disease: Part 2–health effects of omega-3 fatty acids in autoimmune diseases, mental health, and gene expression. *Journal of Medicinal Food*, 8(2), 133–148.

Shahidi, F., & Ambigaipalan, P. (2018). Omega-3 polyunsaturated fatty acids and their health benefits. *Annual Review of Food Science and Technology*, 9, 345–381.

Simopoulos, A. P. (2006a). Evolutionary aspects of diet, the omega-6/ omega-3 ratio and genetic variation: nutritional implications for chronic diseases. *Biomedicine and Pharmacotherapy*, 60(9), 502–507.

Simopoulos, A. P. (2006b). Omega-3 fatty acids in inflammation and autoimmune diseases. *Journal of the American College of Nutrition*, 21(6), 495–505.

Smith, R. E., Finley, J. W., & Leveille, G. A. (1994). Overview of Salatrim, a family of low calorie fats. *Journal of Agriculture and Food Chemistry*, 42(2), 432.

Song, K., Song, I. B., Gu, H. J. et al. (2016). Anti-diabetic effect of fermented milk containing conjugated linoleic acid on type II diabetes mellitus. *Korean Journal for Food Science of Animal Resources*, 36(2), 170–177. doi:10.5851/kosfa.2016.36.2.170.

Stark, K. D. (2008). The percentage of n-3 highly unsaturated fatty acids in total HUFA as biomarker for omega-3 fatty acid status in tissues. *Lipids*, 43(1), 45–53. [PubMed] [Google Scholar]

Subramaniam, S., Fahy, E., Gupta, S., Sud, M., Byrnes, R. W., Cotter, D., Dinasarapu, A. R., & Maurya, M. R. (2011). Bioinformatics and systems biology of the lipidome. *Chemical Reviews*, 111(10), 6452–6490.

Swift, L. L., Hill, J. O., Peters, J. C., & Greene, H. L. (1992). Plasma lipids and lipoproteins during 6 d of maintenance feeding with long chain, medium chain and mixed chain triacylglycerols. *American Journal of Clinical Nutrition*, 56(5), 881.

Terés, S., Barceló-Coblijn, G., Benet, M., Alvarez, R., Bressani, R., Halver, J. E., & Escribá, P. V. (2008). Oleic acid content is responsible for the reduction in blood pressure induced by olive oil. *Proceedings of the National Academy of Sciences of the United States of America*, 105(37), 13811–13816. doi:10.1073/pnas.0807500105.

Thompson, A. K., Minihane, A. M., & Williams, C. M. (2011). Trans fatty acids and weight gain. *International Journal of Obesity*, 35(3), 315.

Uauy, R., & Dangour, A. D. (2006). Nutrition in brain development and aging: Role of essential fatty acids. *Nutrition Reviews*, 64(5 Pt 2), S24–S33.

Umesha, S. S., Manohar, R. S., Indiramma, A. R., Akshitha, S., & Naidu, K. A. (2015). Enrichment of biscuits with microencapsulated omega-3 fatty acid (alpha-linolenic acid) rich Garden cress (Lepidiumsativum) seed oil: Physical, sensory and storage quality characteristics of biscuits. *LWT-Food Science and Technology*, 62(1), 654–661.

Valentinotti, S., Armanet, L., & Porret, J. (2006). Encapsulated polyunsaturated fatty acids WO2006/067647.

Verardo, V., Ferioli, F., Riciputi, Y., Iafelice, G., Marconi, E., & Caboni, M. F. (2009). Evaluation of lipid oxidation in spaghetti pasta enriched with long chain n– 3 polyunsaturated fatty acids under different storage conditions. *Food Chemistry*, 114(2), 472–477.

Wang, C., Chung, M., Balk, E., et al. (2006). n-3 Fatty acids from fish or fish-oil supplements, but not -linolenic acid, benefit cardiovascular disease outcomes in primary and secondary prevention studies: A systematic review. *American Journal of Clinical Nutrition*, 83, 5–17.

Watkins, B. A., Li, Y., Romsos, D. R., Hoffman, W. E., Allen, K. G. D., & Seifert, M. F. (2003). CLA and bone modeling in rats. In: J. Sebedio, W. W. Christie and R. Adolf, editors. *Advances in Conjugated Linoleic Acid Research*, Vol. 2. AOCS Press: Champaign, IL, pp. 218–250.

Wendell, S. G., Baffi, C., & Holguin, F. (2014). Fatty acids, inflammation, and asthma. *Journal of Allergy and Clinical Immunology*, 133(5), 1255–1264.

Willis, W. M., Lencki, R. W., & Marangoni, A. G. (1998). Lipid modification strategies in the production of nutritionally functional fats and oils. *Critical Reviews in Food Science and Nutrition*, 38(8), 639–674.

Willis, W. M., & Marangoni, A. G. (1999). Assessment of lipase and chemically catalysed lipid modification strategies for the production of structured lipids. *Journal of the American Oil Chemists' Society*, 76(3), 443–450.

Xu, X. B. (2000). Production of specific-structured triacylglycerols by lipase-catalyzed reactions: A review. *European Journal of Lipid Science and Technology*, 102(4), 287–303.

Yang, T., Xu, X., He, C., & Li, L. (2003). Lipase-catalyzed modification of lard to produce human milk fat substitutes. *Food Chemistry*, 80(4), 473–481.

Yang, L. G., Song, Z. X., Yin H. et al. (2016). Low n-6/n-3 PUFA ratio improves lipid metabolism, inflammation, oxidative stress and endothelial function in rats using plant oils as n-3 fatty acid source. *Lipids*, 51(1): 49–59.

Ye, A., Cui, J., Taneja, A., Zhu, X., & Singh, H. (2009). Evaluation of processed cheese fortified with fish oil emulsion. *Food Research International*, 42(8), 1093–1098.
Yep, Y. L., Li, D., Mann, N. J., Bode, O., & Sinclair, A. J. (2002). Bread enriched with microencapsulated tuna oil increases plasma docosahexaenoic acid and total omega-3 fatty acids in humans. *Asia Pacific Journal of Clinical Nutrition*, 11(4), 285–291.
Zam, W. (2015). Structured lipids: Methods of production, commercial products and nutraceutical characteristics. *Progress in Nutrition*, 17(3), 198–213.

INDEX